5.95

REWARD

2005

Grammar and Vocabulary Workbook

Simon Greenall

MACMILLAN
HEINEMANN
English Language Teaching

Macmillan Heinemann English Language Teaching
Between Towns Road, Oxford OX4 3PP
A division of Macmillan Publishers Limited
Companies and representatives throughour the world

ISBN 0 435 24271 7 (with key)

ISBN 0 333 74253 2 (without key)

First published 1998

Designed by Sarah Nicholson

Printed and bound in Spain by Mateu Cromo, S.A. Pinto (Madrid)

2003 2002 2001 2000
12 11 10 9 8 7 6 5 4

Contents

Lessons 1–5

VOCABULARY

1 Write the questions.

1 _____ ?
  Smith.

2 _____ ?
  S-M-I-T-H.

3 _____ ?
  Twenty-four.

4 _____ ?
  No, I'm single.

5 _____ ?
  6, Kingston Road, Dover.

6 _____ ?
  I'm British.

7 _____ ?
  Football – I love watching football.

8 _____ ?
  In a restaurant. I'm a cook.

2 Match a word on the left with a word or phrase on the right. Use each phrase only once.

1	read	a the activity
2	complete	b notes
3	ask	c the correct answer
4	do	d the passage
5	explain	e the chart
6	repeat	f the general meaning
7	tick	g any new words in the dictionary
8	make	h the word you hear
9	underline	i another student
10	look up	j the sentence slowly

3 Choose the correct word to fill the gap.

1 The _____ for the wedding took six months.
  (preparations/celebrations)

2 Barry bought a ring for Anna to mark their _____.
  (entertainment/engagement)

3 They decided to have a _____ wedding in church.
  (ceremonial/traditional)

4 The _____ wore a long white dress and the _____ wore pink.
  (bride/groom; bridesmaids/matchmaker)

5 A hundred people came to the _____ after the ceremony.
  (reception/registry office)

6 All the _____ brought a present.
  (gifts/guests)

7 Everybody got a piece of _____ cake. (marriage/wedding)

8 The couple are spending their _____ in Florida.
  (horoscope/honeymoon)

4 Put one of the words from the list in each of the gaps.

crowds booking office traffic catch
platform fare connection suitcases
round trip announcement reservation
taxi rank

Every Friday evening I drive to the station and get the train to London to meet my boyfriend. I leave the flat at six o'clock. The (1)_____ is terrible at that time and the journey takes twenty minutes. I have to get a train to Cambridge first and then wait for a (2)_____. I listen to the (3)_____ about which (4)_____ to go to and then I try and find a seat. If I haven't made a (5)_____ I sometimes have to stand all the way. In London there are usually (6)_____ of tourists with huge (7)_____ waiting for tickets in front of the (8)_____. I squeeze past them and join the queue at the (9)_____ and wait for a taxi. If I'm at the back of the queue I feel impatient and nervous. I really hate waiting. My boyfriend always says, 'Why don't you (10)_____ the bus?' but that takes even longer. The taxi (11)_____ from the station is £4.50. Altogether the (12)_____ costs me about £30. Should I change my job and work in London or should I get a new boyfriend?

5 Put the words under the correct heading.

opera DIY reading folk cricket
train spotting tennis shopping running
lyrics swimming bird-watching rehearsal
classical walking

Sport	**Music**
_____	_____
_____	_____
_____	_____
_____	_____
_____	_____

Hobbies

_____
_____
_____
_____
_____

6 Match a sentence in A with a sentence in B.

A

1 He's a couch potato. [c]
2 She plays in an orchestra. [ ]
3 I love football. [ ]
4 My parents adore entertaining. [ ]
5 He's always busy in the garden. [ ]
6 She hates watching television. [ ]
7 I think fishing is boring. [ ]

B

a Her favourite instrument is the cello.
b She likes doing odd jobs instead.
c He never does any sport.
d I never catch any fish.
e I go to all the home matches.
f He's cutting the grass now.
g Personally, I enjoy going to restaurants.

7 Rearrange the letters in the following words. Then put them in the sentences below.

egtar leduarfd lal htrig fricreti
grinob texngici

1 'The film was so dull.' – 'Yes, I agree, it was very _____.'

2 'Did you have fun at the party?' – 'Yes, thank you, I had a _____ time.'

3 'Was your journey OK?' – 'Well, there was a short delay but then it was _____.'

4 'Isn't this music brilliant? I think Purple Veil are a _____ band.'

5 'My homework was awful. The teacher said I made some _____ mistakes.'

6 'Bird-watching is a very relaxing hobby but I like _____ sports like football, too.'

8 Complete the sentences.

1 A _surprising_ piece of news is one that you didn't expect to hear.

2 A _b_____ film is one that you don't enjoy watching.

3 If you feel sad and lonely, you feel _d_____.

4 You get _e_____ at a football match when your team is winning.

5 It must be very _f_____ to see a ghost.

6 Sitting by the river on a hot day is very _r_____.

7 He was _a_____ because the train was cancelled.

8 I was _s_____ to hear about the accident.

9 My exam results were _d_____. I only got 55%.

10 She was _t_____ when she won first prize of a trip to the moon.

9 Complete the sentences with the best word a, b, c or d.

1 What's making that funny _____? Can you hear it?
a music b horn c noise d paste

2 How much is the _____ to the airport?
a fare b fear c folk d funk

3 Red is a colour that will bring you _____ luck.
a green b gold c happy d good

4 The mosquitos were very _____.
a moving b silly c depressing d irritating

5 I'm afraid I've run _____ milk.
a into b out of c off with d over

6 They're spending their _____ in Los Angeles.
a free time b shopping c vacation d reception

7 She often stays _____ until 3 am.
a out b off c over d on

8 The tennis _____ was cancelled because of the weather.
a game b club c match d play

9 I don't _____ reggae but I hate techno.
a mind b stand c mean d adore

10 I don't want to _____ my engagement ring.
a leave b miss c wish d lose

GRAMMAR

1 Choose the correct verb form.

1 She *plays/is playing* the cello every day.

2 They *collect/are collecting* the leaves in the autumn.

3 The crowds *shout/are shouting* because the match is over.

4 We always *have/are having* dinner with my husband's family on Sunday.

5 I *meet/am meeting* my friend at the cinema next Saturday.

6 He *works/is working* as a cab driver at the moment.

7 At a Hindu wedding the bride *wears/is wearing* a red sari.

8 Many people *believe/are believing* that red brings good luck.

2 Put the words in the right order and make questions.

1 your / with / sisters / does / live / mother / her / ?

_____

_____

2 how / see / brother / you / often / do / your / ?

_____

_____

3 platform / the / leave / London / which / 6.15 / does / to / train / from / ?

_____

_____

4 give / bride / you / what / groom / did / and / to / the / ?

_____

_____

5 is / guitar / playing / band / who / rock / the / this / in / ?

_____

_____

3 Complete the sentences with a suitable question tag.

1 Your wife comes from Budapest, _____ ?

2 He doesn't play cricket at school, _____ ?

3 They're spending their honeymoon in Niagara Falls, _____ ?

4 You don't like decorating, _____ ?

5 My brother's got a fast car, _____ ?

6 That rock band play very exciting music, _____ ?

7 We don't often go to the beach at the weekend, _____ ?

4 Make true sentences about yourself using the phrases below and *never, sometimes, usually, often, always*.

get up at
play football at the weekend
plant bulbs in the garden
make mistakes
spend my holidays in
have dinner with
go window shopping
have lunch at
listen to pop music
go train spotting

1 _____

_____

2 _____

_____

3 _____

_____

4 _____

_____

5 _____

_____

6 _____

_____

7 _____

_____

8 _____

_____

9 _____

_____

10 _____

_____

5 Rewrite the sentences using the word in brackets.

1 I book my ticket. Then I check which platform the train is leaving from. (after)

After I book my ticket, I check which platform the train is leaving from.

2 They have weeks of rehearsals. Then they perform the music in front of hundreds of people. (before)

3 We were engaged for six months. Then we got married. (before)

4 They were married for twenty years. Then they suddenly got divorced. (after)

5 He spends his holidays doing odd jobs in the house. (during)

6 We had dinner. The telephone rang three times. (during)

7 He plays tennis every Sunday from 2.30 to 5.30. (for)

8 She put the television on at nine o' clock and then at ten o'clock she went to bed. (for/before)

6 What do you say in the following situations?

1 You have forgotten your pen. Your friend has several in her bag.

2 Your teacher asked you a question but you didn't hear.

3 You hear your teacher say a word which you want to record in your notebook.

4 Your friend uses a word in English which you don't understand.

5 Your teacher is speaking too quickly.

6 You are trying to describe something in English but you don't know the right word to use.

7 You want the person next to you on the train to tell you when you arrive at Sheffield.

7 Look at the chart, then make sentences about the people.

	Julio	**Gunther**	**Claude**	**Diana**	**Eva**	**Ingrid**
detest	weed the garden				play cards	
hate			decorate			
can't stand		play cricket		watch TV		
don't mind		play rugby			make cakes	
like			go bird-watching			entertain
love	play the cello					
adore				go window shopping		go to nightclubs

1 Julio   *detests weeding the garden but he loves playing the cello.*

2 Gunther   _____

3 Claude   _____

4 Diana   _____

5 Eva   _____

6 Ingrid   _____

8 Look again at the chart above and write similar sentences about yourself.

_____

_____

_____

_____

WRITING

Write a letter to a penfriend in Britain telling him or her what you and your friends do in your free time. Write about sports, hobbies, music and entertainment. Ask your penfriend about his or her free time and about his or her likes and dislikes. Include your address and the date and choose an appropriate way to end the letter.

Lessons 6–10

VOCABULARY

1 Circle the odd-word-out.

1 seaweed shells sand ice cream spices
2 explorer neighbour teacher waiter actress
3 banana bean beetroot cabbage carrot
4 cold chilly moody cool freezing
5 reliability honesty healthy sincerity

2 Choose the correct word to fill the gap.

1 When I lost my map I became _____ and took the wrong turning.
(foolish/confused)

2 I stayed with a very _____ family who made me feel at home.
(hospitable/fashionable)

3 He spoke to his wife in a strange language. I assumed they were _____ .
(strangers/foreigners)

4 At last I was _____ in the house and could do exactly what I liked.
(lonely/alone)

5 She got to know him as a _____ when she worked at the hospital.
(colleague/customer)

6 They watched a group of _____ walking down the road in uniform.
(soldiers/shoulders)

7 I usually carry my _____ in my shirt pocket.
(bag/wallet)

3 Find the opposites.

~~solid~~ freezing sweet dishonest savoury
even grown-up miserable ~~liquid~~ stale odd
healthy boiling honest happy boring fresh
child interesting ill

1 *solid – liquid* _____
2 _____
3 _____
4 _____
5 _____
6 _____
7 _____
8 _____
9 _____
10 _____

4 Add the prefix *un-*, *im-* or *in-* to each of the following words to make a word with the opposite meaning.

certain _____
clear _____
complete _____
correct _____
expensive _____
friendly _____
happy _____
important _____
kind _____
patient _____
perfect _____
possible _____

5 Read the sentences below and put the correct words in the puzzle.

1 |☐| — — — — — —
2 |☐| — — — — — — —
3 |☐| — — — — —
4 |☐| — — — — — — — —
5 |☐| — — — — — — —
6 |☐| — — — — — — — —
7 |☐| — — — — — — —

1 This is something you give to someone.

2 This describes someone who is not shy but friendly and likes to meet new people.

3 When I feel this very strongly I can say I am delighted.

4 The word for this comes from Aztec and it makes us feel we are falling in love.

5 You feel this when you hope for good things in the future.

6 He was wearing yellow shorts, an orange shirt and a big fur hat – he looked ....

7 This is someone who lives near your house.

Now read down the first vertical column to find a word for something that's good to eat. _____

6 Match the two parts of the sentences.

1 I am looking forward ☐ *e*
2 Deborah has fallen in love ☐
3 We soon got to know ☐
4 He's trying to get in touch ☐
5 The story of the film takes place ☐
6 We would like to complain ☐
7 We share the same tastes ☐

a our friendly pizza delivery boy.

b about our anti-social neighbours.

c in books and music.

d with a good-looking young actor called Scott.

e to a meal in an expensive French restaurant.

f with a company that exports cocoa.

g in the neighbourhood where we used to live.

7 Complete the sentences.

1 It was very embarrassing. I didn't realise we were supposed to go *Dutch*_____ and I didn't have any money with me.

2 When I was a student I ate bread, cabbage, bananas and things like that. I could never *a*_____ to buy meat.

3 You can find your way if you look for the tower on the top of the hill. It's the best *l*_____ we've got in this town.

4 The best place to buy fresh vegetables is at the *g*_____ .

5 I had nowhere to sleep, all the hotels were full and I didn't understand what people were saying to me. I was beginning to *p*_____ when suddenly I heard someone say 'Hello!'

6 After he won all that money, he couldn't decide if he should spend it all or *i*_____ it.

7 When the baby was born, it was hard to *c*_____ with all the sleepless nights.

8 Rearrange the letters in the following words. Then use the words to complete the sentences below.

ogod   liongib   uehg   gyrna   risoluhia
ezgrnefi   gleedthid   ylsli   baslemeir   trigsennite

1 The pizzas he served us weren't just big, they were _____ .

2 They spent a _____ evening at the pub telling each other jokes and funny stories.

3 Put your hat and gloves on before you go out – it's _____ .

4 I found his talk _____ and the bit about the history of the diamond mines was fascinating.

5 What an excellent performance by the soloist! Unfortunately the orchestra wasn't really very _____ .

6 She always eats chocolate when she's feeling _____ . It doesn't make her happy but it stops her smoking.

7 Sit here under the trees where it's cool. Out there by the pool it's _____ .

8 My father was furious when I told him. I've never seen him so_____ .

9 I know this sounds ridiculous but I've forgotten my telephone number. Do you ever do _____ things like that?

10 Was he pleased when you found that old sporting cup he won for golf? - To tell you the truth, I think he was _____ .

9 Choose the correct spelling.

1 It happened last *weak/week*.

2 The room was small and rather *plain/plane* with a brown carpet.

3 Eat your dinner or you'll be *angry/hungry*.

4 *Prizes/Prices* are going up in the shops.

5 I mustn't eat any more chocolate. My *waste/waist* is disappearing!

6 We spent the morning on the *beach/beech*.

GRAMMAR

1 Read the story and write the correct form of the verb in italics.

Yesterday I (1) *carry* _____ some heavy shopping from the car to my front door when the telephone (2) *start* _____ to ring inside the house. I put the bags on the ground and took out my key. As I (3) *open* _____ the door, I (4) *hear* _____ a noise. The car alarm was going off. I (5) *rush* _____ back to the car, (6) *open* _____ the door and (7) *switch* _____ the alarm off. A dog (8) *walk* _____ down the street and it (9) *smell* _____ the chicken in

my shopping bag. It (10) *run* _____ across the garden and (11) *pull* _____ the chicken out of the bag when I (12) *catch* _____ it. When I turned round, my neighbour (13) *stand* _____ in front of my house with my key in her hand. 'I saw it in the lock,' she said. 'You must be careful.' I (14) *carry* _____ my shopping into the house and then (15) *walk* _____ back to the car. It was locked and the key (16) *lie* _____ on the front seat. I had to phone a garage and ask someone to come and help. Two hours later, I (17) *eat* _____ an incomplete meal of potatoes and carrots when the phone (18) *ring* _____. It was my friend. 'I (19) *phone* _____ you several times earlier this afternoon without getting any answer. Whatever (20) *you/do* _____ ?'

2 Put the words in the correct order to make sentences. One sentence is false. Which one?

1 black be television and used programmes white to in

_____

_____

2 one to used there channel only be

_____

_____

3 special a invite their watch people to neighbours programme would round

_____

_____

4 drinking use law there about and didn't to driving be a

_____

_____

5 control go Scotland you through England to between used passport and

_____

_____

6 each walk to several children day school would miles

_____

_____

7 used every sterling there twenty be in pound to shillings

_____

_____

3 Join the sentences using the words *as, as soon as, just as, until, when* and *while*.

1 Ben's mother announced the start of the treasure hunt. The children ran into the garden.

_____

_____

2 I noticed my friend on the other side of the street. I waved.

_____

_____

3 She swallowed her seventh chocolate. She read the number of calories on the side of the box.

_____

_____

4 The plane was landing. She recognised the man sitting across the aisle.

_____

_____

5 He was walking along the road. He whistled a tune to himself softly.

_____

_____

6 I was waiting in the queue for tickets. Someone tapped me on the shoulder.

_____

_____

7 They walked off the boat onto dry land. They were still feeling sick.

_____

_____

8 I searched the whole beach. I finally found Jo playing with some shells near the water.

_____

_____

_____

9 She was training to be a teacher. A film director spotted her.

_____

_____

_____

10 He was looking puzzled. I reminded him that we had met at a party.

_____

_____

_____

4 Read the sentences 1–7. Choose one extra piece of information (a–g) to add to each sentence. Make the necessary grammatical changes.

1 Mother Theresa spent her whole life looking after sick people.
2 Spices grow mainly in Eastern countries.
3 Bilbao is on the north coast of Spain.
4 Jane Austen wrote stories about life in quiet country villages.
5 If you like old postcards, go to one of the junk shops.
6 My uncle sometimes phones me at home and tells me he's a policeman!
7 Edward Whymper, an English explorer, was the first man to climb mountains in Ecuador.

a She had a sister called Cassandra.
b He has a sense of humour.
c You can find them in every town in Britain.
d You can still find streets named after him there.
e They improve the taste of many dishes.
f She lived in Calcutta.
g They built a huge new museum there.

1 *Mother Theresa, who lived in Calcutta, spent her whole life looking after sick people.*

2 _____

_____

_____

_____

3 _____

_____

_____

_____

4 _____

_____

_____

_____

5 _____

_____

_____

_____

6 _____

_____

_____

_____

7 _____

_____

_____

_____

5 What do you say in the following situations? Use the words in brackets.

1 Tell somebody that you received a box of chocolates from your neighbour. (give)

My neighbour gave me a box of chocolates.

2 Thank a person who has carried a cup of tea to you in bed. (bring)

3 Ask someone for the salt and pepper. (pass)

4 You went for a job interview and you were successful. Tell a friend. (offer)

5 Ask somebody to confirm that they borrowed £5 from you a week ago. (lend)

6 A friend asks you, 'What is a suitable present for me to give to my aunt in hospital when I go to see her?' Suggest fruit. (take)

7 Your friends in Australia have a new baby. You want to see what he looks like. (send)

8 You were unable to return a friend's phone message because you did not know her number. Explain your reason to her. (leave)

9 Tell somebody that you and your partner spent £300 on a week in Majorca. (cost)

10 Introduce Tina White, the poet, to an audience who have come to hear her work. (read)

6 Describe the progress made by a student who has just completed a year in England with an English family and who now enjoys living in Britain.

1 When I first came to Britain,

2 After a while,

3 Eventually

4 Now

WRITING

1 Write the dialogue in a restaurant between the waiter and a customer.

The waiter politely tells the customer that smoking is not allowed in this part of the restaurant. He asks him to move to another table.

Waiter: _____

_____

_____

The customer apologises because he did not read the sign.

Customer: _____

The waiter replies.

Waiter: _____

The customer asks the waiter for the bill.

Customer: _____

The waiter brings it.

Waiter: _____

The customer then notices a mistake in the bill.

Customer: _____

The waiter apologises.

Waiter: _____

The customer accepts his apology.

Customer: _____

2 Describe the memories you have of family parties, for example, Christmas, New Year or another festival. Which members of your family were there? Write down some interesting information about them. What happened? Did you all enjoy yourselves? Why (not)?

Lessons 11–15

VOCABULARY

1 Read the statements and ask questions using *Have you …?*

1 The postman has just given me today's mail.

   Have you opened it yet?

2 They sent the bill three weeks ago.

   _____

3 Ten people wrote to me last week, asking questions about the new sports centre.

   _____

4 The children finish school at 3.15. It's now 3.30.

   _____

5 The garage left a message to say the car is ready.

   _____

6 She told me the TV wasn't working yesterday.

   _____

7 I'm sending some books to Aunt Julia for her birthday.

   _____

8 I've read all of last week's newspapers.

   _____

9 She left me a shopping list before she went out.

   _____

2 Complete each sentence with one of the following words.

fictional   criminal   evil   typical   popular
frustration   success

1 Hannibal Lecter represents pure _____ in the film *The Silence of the Lambs*.

2 *Four Weddings and a Funeral* was a very _____ film.

3 The most famous _____ detective is Sherlock Holmes.

4 Sometimes the 'hero' of a film can be a _____.

5 A _____ film by Stephen Spielberg shows some ordinary people in a fantastic setting.

6 Nobody had predicted the _____ of *Babe*, a film about a pig.

7 It's a film that deals with the _____ of deaf people who are trying to communicate with people who can hear.

3 Choose the correct word to make a compound noun.

1 Working on an oil *rig/ring* is difficult but well paid.

2 That's a pretty wrist *clock/watch* you're wearing.

3 The concert *house/hall* where the orchestra will play is in the centre of town.

4 Pass me the remote *control/cartoon*. I want to change the channel.

5 I saw some wonderful paintings in the art *gallery/frame*.

6 I prefer writing with a *mountain/fountain* pen.

7 Is this picnic *minister/basket* big enough for all our food and wine?

Now match each of the words you didn't use with one of the following.

coffee   strip   engagement   prime   alarm
photo   bike

_____

_____

_____

_____

_____

4 Circle the odd-word-out.

1	bomb	bullet	missile	crew
2	player	border	lawyer	officer
3	gale	storm	weapon	rain
4	firefighters	terrorists	hostages	hijackers
5	battle	victory	navy	currency
6	president	kidnap	judge	soldier

5 Change one word in each sentence so that it makes sense. Choose from these words.

market   square   river   entertainment   cathedral
population   architecture   quay

1 Boats leave the ~~lane~~ every hour on trips round the island.

   *quay*_____

2 The museum had lots of fresh fruit and bright clothes for sale.

   _____

3 The skyscraper was a very tall stone building and was said to be over 700 years old. Inside it was cool and quiet.

   _____

4 Most of the shops are on the park in the centre of town. Do you know where I mean? There are traffic lights at each corner.

   _____

5 The old-fashioned industry makes the town look very picturesque.

   _____

6 The district is famous for its cosmopolitan cemetery and lively atmosphere.

   _____

7 You can find a wide range of transport in this theatre – plays, opera, concerts and comedies.

   _____

8 The palace is unique because it was built on the banks of a castle and can only be visited by people who have walked over the bridge.

   _____

6 Complete the sentences using an adjective from list A and a preposition from list B.

A afraid   allergic   capable   pleased   fond
   similar   bored   proud

B of   to   with

1 She became _____ _____ sitting still for so long.

2 He is no good as a manager. He isn't _____ _____ making decisions.

3 She is _____ _____ chocolate – it makes her sick.

4 He was very _____ _____
   himself for thinking of such a clever trick.

5 They are very _____ _____
   their daughter – she has just won an
   important prize for music.

6 Your jacket is very _____
   _____ mine – I can't see any difference
   between them.

7 There is only one thing I am
   _____ _____ – spiders!

8 She doesn't know which of her children to
   go on holiday with – she is very
   _____ _____ all
   of them.

7 Put the adjectives in the right order to
describe each noun. Write the phrase.

1 a table
   (antique/beautiful/mahogany)

   _____

2 a violin
   (heavy/Italian/old)

   _____

3 a necklace
   (Cartier/diamond/brand-new)

   _____

4 a sofa
   (green/large/Victorian)

   _____

5 a vase
   (blue/porcelain/lovely)

   _____

8 Rearrange the letters in the following
words and put them in the sentences
below.

lil   viale   loean   darye   rosyr   agld

1 I was _____ to hear from you after
   such a long time.

2 My brother has had an accident but
   fortunately he is still _____.

3 When you heard strange noises in the
   house were you quite _____ ?

4 She needed an operation. It seems she has
   been very _____ for some time now.

5 The athlete lost points because he was not
   _____ when the starter blew the
   whistle.

6 I am _____ to hear that you failed
   your exam. Good luck next time!

9 Complete the sentences with the best
word, a, b, c or d.

1 The _____ was packed tonight, as
   30,000 fans watched an exciting match
   between Italy and Holland.

   a quarter b suburbs c stadium d wing

2 A plane was trying to land at Heathrow
   Airport when lightning struck a power
   cable and the _____ runway
   was plunged into darkness.

   a attractive   b illuminated   c memorable
   d cosmopolitan

3 There was a strike on the city's transport
   services today and the only method of
   crossing the centre was on _____ .

   a tram b foot c team d quay

4 Some people make a strange choice of
   animal to keep as a _____ .

   a frog b dog c rabbit d pet

5 The journalist was awarded the Journalist of the Year prize for the way she reported wars, international treaties, the single currency and other foreign _____ .

a attacks   b affairs   c interests   d stories

6 _____ are free to explore the old quarter without fear of meeting any traffic.

a Precedents   b Prints   c Pedestrians
d Tourists on bicycles

7 This collection of books and prints is _____, as the only other example was destroyed by fire.

a rare   b unique   c antique   d romantic

GRAMMAR

1 Complete the sentences below with a suitable possessive pronoun or adjective.

1 You've got a book like this, haven't you? Is this _____ book or _____ ?

2 My sister and her husband have both got cars. _____ is second-hand but _____ is new.

3 My bicycle was stolen last week. Look at that bicycle over there. I'm sure that's _____.

4 They're only 20 and they've already bought _____ own house. We didn't buy _____ until we were nearly 30.

5 I bought them a ticket for the lottery. Then they phoned me to say _____ was the winning ticket.

6 He forgot _____ keys, but luckily his wife was with him and they were able to use _____ .

2 Petra is going on holiday today. Look at her list. What has she already done? What hasn't she done yet? Write sentences.

1	Make sandwiches for the journey ✓
2	Pack the suitcases
3	Put petrol in the car ✓
4	Turn off the electricity
5	Water the plants ✓
6	Buy some maps ✓
7	Tidy the house
8	Phone parents ✓
9	Take out the rubbish
10	Buy suncream

1 *She's already made sandwiches for the journey.*

2 *She hasn't packed the suitcases yet.*

3 _____

_____

4 _____

_____

5 _____

_____

6 _____

_____

7 _____

_____

_____

8 _____

_____

9 _____

_____

10 _____

_____

3 Write sentences in the present perfect continuous with *since* or *for*.

1 We started playing tennis at two o'clock. It's four o'clock now and we're still on the tennis court.

We've been playing tennis for two hours.

2 She learnt to play the violin five years ago. She still loves playing it.

_____

_____

3 He passed his driving test when he was eighteen. He's thirty-three now.

_____

_____

4 Her boyfriend phoned at seven o'clock. They haven't stopped talking.

_____

_____

5 I felt miserable when I woke up this morning. My mood hasn't changed all day.

_____

_____

4 Answer the questions using two of these words each time.

run   live   football   ~~love story~~   bus   play   rain   ~~cry~~   Germany   pain   shoulder   walk

1 Why are her eyes red?

She's been crying over a love story.

2 Why are you going to the doctor's?

_____

3 Why are your clothes dirty?

_____

4 Why are they out of breath?

_____

5 How did you get so wet?

_____

6 You speak German very well now. How did you learn so quickly?

_____

5 Write out the dialogue below. Make the necessary grammatical changes.

A Hello, Jan. I/not/see/you at the line-dancing classes/a few weeks.

_____

_____

B Hello, Linda. I/not/have/enough time, I'm afraid.

_____

_____

A What/you/do?

_____

B I/study/for my English Literature exam.

_____

_____

A Oh really? How many books/you/study?

_____

_____

B I/already read/three novels by Jane Austen and I/read/Shakespeare's *King Lear*, but I'm finding it quite difficult.

_____

_____

_____

_____

A you/read/Hamlet?

_____

B No, but I/see/the film.

_____

A you/read/any poetry?

_____

B Yes, in fact, I/try/to learn some poems by heart. I/saying/them to myself in the bus on the way to work.

_____

_____

_____

A I like writing poems. I/write/them/since/I/leave/school, but I/not/finish/one yet.

_____

_____

_____

6 Make comparisons using the word in brackets and give your opinion.

1 Which is more difficult, English or Chinese? (easy)

_____

_____

2 There are fewer cars in the centre of Rome than in Athens. (pollution)

_____

_____

3 My little girl screamed when we went to see *Jurassic Park*. Should we go to see *ET*? (frightening)

_____

_____

4 Why don't they show Liverpool in tourist brochures as often as Oxford? (picturesque)

_____

_____

5 Why does Bill Gates have a bigger house than my father has? (money)

_____

_____

6 Does London have as many skyscrapers as New York? (many)

_____

_____

WRITING

Describe yourself and your best friend and say how long you have known each other. Decide what similarities there are between you, for example, your age, your family background, your interests, and then write about the differences. Explain why you became friends and why you like to spend time together.

Lessons 16–20

VOCABULARY

1 Find 18 names for colours and shades in the word box. The words go horizontally, vertically and diagonally and you can read forwards or backwards. When you have found all the words, write down the letters you did not need. Rearrange them to make two more colour words.

K	U	S	W	H	I	T	E	O	L
H	C	H	O	C	O	L	A	T	E
A	A	A	L	M	A	E	R	C	G
K	N	O	L	I	V	E	Q	R	N
I	A	R	E	B	L	U	E	I	A
C	R	U	Y	E	R	G	E	M	R
R	Y	T	E	I	E	O	I	S	O
N	E	E	R	G	D	A	W	O	T
S	T	O	N	E	Y	V	A	N	S

1 _____   10 _____

2 _____   11 _____

3 _____   12 _____

4 _____   13 _____

5 _____   14 _____

6 _____   15 _____

7 _____   16 _____

8 _____   17 _____

9 _____   18 _____

Extra words:

_____   _____

2 Complete the sentences.

1 If you have long hair you can choose to wear it in a p_____ .

2 I didn't know he had a c_____ c_____ until he took off his baseball cap.

3 Are you going to grow a beard or is that just designer s_____ ?

4 When she tossed her hair back I could see her silver e_____ .

5 First he put on his b_____ s_____ and then his jeans.

6 Both men and women can wear f_____- f_____ instead of shoes in the summer.

7 Are you sure you can walk in those h_____ h_____ ?

3 Answer the questions about people's jobs.

1 Who wears a uniform?

_____

_____

2 Who works outside?

_____

_____

3 Who works in a hospital?

_____

_____

4 Who works in a studio?

_____

_____

5 Who works in an office?

_____

_____

4 Complete the sentences with the best word, a, b, c or d.

1 The model, who was wearing clothes for skiing, walked _____ across the car park and climbed into a brand-new sports car.
a daintily b lively c conspicuously
d charmingly

2 The patient complained about the _____.
a screen b draught c brush d stove

3 Our _____ were so heavy we had to rest every hour on our way up the mountain.
a equipment b packages c scallops
d backpacks

4 I don't know what time I'll arrive. It depends _____ the traffic.
a on b off c down d up to

5 Inside the church was a marble _____.
a pilgrim b souvenir c incense d statue

6 The _____who had conducted the service led the way down the steps of the cathedral.
a attendant b priest c saint d ghost

5 There is a mistake in each of these sentences. Cross out the wrong word and write the correct word underneath.

1 We travelled round the city ~~with~~ bicycle.

   by _____

2 Pat wanted to make some sightseeing first.

   _____

3 Gloria nibbled her orange juice through a straw.

   _____

4 He was showing on his new uniform to his friends.

   _____

5 She offered to lend me her jacket-knife.

   _____

6 She was dressed in head-to-heel silk.

   _____

7 He spent the whole weekend by his own.

   _____

6 Find the opposites. Match a word from A and a word from B.

A	B
cool	serious
fashionable	fair-minded
flamboyant	deceitful
fun	naïve
honest	self-assured
lazy	emotional
prejudiced	middle-aged
sensitive	hard-working
shy	reserved
sophisticated	old-fashioned
young	thick-skinned

7 Make compound adjectives by matching a word from A and a word from B.

A	B
cool	behaved
easy	famous
good	going
left	handed
short	headed
short	hearted
soft	humoured
strong	sighted
well	tempered
world	willed

Put one of the compound adjectives into each sentence.

1 You need to be _____ in a crisis and keep everything under control.

2 The Irish are the most _____ people I know – they always have time to stop and chat.

3 They became _____ after their album went straight to Number One in the pop charts.

4 The children were very _____ - _____ during the concert. They didn't make a noise or move about.

5 I couldn't persuade her to change her mind. She's determined to go off on her own round the world. She's always been very _____ .

6 If you want someone to take pity on you and help you look after your pets while you're away, you should ask Josephine – she's the _____ member of the family.

7 I started wearing glasses when I was at school. The doctor said I was very

_____ .

8 I like your Uncle Bill. He's a very _____ man and never gets angry or impatient.

9 Why can't they put the switch on the other side for _____ people like me?

10 He's become very _____ in his old age. I think it's the frustration of not being independent.

8 Complete the sentences with multi-part verbs.

1 'Wendy's coming round to baby-sit.' 'Oh good. The children *get*___ *on*___ *with*_ her very well."

2 I applied for a job and they've sent me a form so that I can give them the information they need. I've never seen a form like this before. Could you help me *f*_____ _____ _____ , please?

3 'I can hardly hear the radio.' 'Oh, sorry, I *t*_____ _____ _____ so it wouldn't wake the baby.'

4 'I've *p*_____ _____ my dental appointment until next week.'

5 Your bag is on the floor. Do you want me to *p*_____ _____ _____ ?

6 His grandmother *b*_____ _____ _____ after his parents died in a plane crash.

7 Please *t*_____ _____ your shoes before you come into the house. They're very muddy.

8 She was always very shy at school. The other children would be playing games and she never asked if she could *j*_____

_____ .

9 What time does your plane *t*_____ _____ ?

10 Another chocolate! I can see you've decided to *g*_____ _____ your diet.

GRAMMAR

1 A teacher is reminding the students about what to do in the examination room. Complete the sentences with *must, mustn't, should, shouldn't, are supposed to, are not allowed to.*

1 You _____ remember to write your name at the top of the paper.

2 You _____ attempt to answer all the questions.

3 You _____ write with a fountain pen, not a biro, and definitely not a pencil.

4 You _____ speak to anyone during the exam.

5 You _____ write notes on the examination paper. Use a piece of scrap paper.

6 You _____ leave the room before 12.00.

2 Explain what is meant by the following, using the words in brackets.

1 PASSPORT CONTROL   (must)

   You must show your passport as

   you go through the gate.

2 NO SMOKING (allowed)

   _____

   _____

3 SPEED LIMIT 30 MPH   (must)

   _____

   _____

4 HELMETS MUST BE WORN   (have to)

   _____

   _____

5 NO VISITORS AFTER 10 PM   (allowed)

   _____

   _____

6 DRY-CLEANING RECOMMENDED
                              (should)

   _____

   _____

7 SLOW – CHILDREN PLAYING   (must)

   _____

   _____

8 NO CAMERAS IN THE AUDITORIUM
                              (supposed)

   _____

   _____

3 Look at the signs again in Exercise 2. Where do you think you would see them?

1 *This must be at a border crossing,*

   an airport or a port.

2 _____

   _____

3 _____

   _____

4 _____

   _____

5 _____

   _____

6 _____

   _____

7 _____

   _____

8 _____

   _____

4 Read what happened in a museum. You are the museum attendant. What did you say?

1 A man took a photograph of a painting. (allowed)

You are not allowed to take

photographs, sir.

2 A woman picked up a small statue. (touch)

_____

_____

3 A young man was carrying a shopping bag on his shoulder in the main gallery. (leave/cloakroom)

_____

_____

4 Two boys sat down and took out sandwiches and cans of Coke. (supposed/food and drink)

_____

_____

5 A young woman's mobile phone rang and she answered it. (allowed)

_____

_____

6 A man with a camera and a bag of postcards walked past a sign that said *Tourists are asked to contribute £3.00 for each visit.* (supposed/entrance fee)

_____

_____

7 A guide was talking too loudly to her group. (quietly)

_____

_____

5 Read what two different people say when they talk about their career plans. A is very confident about becoming successful. B is not certain about her plans. Put appropriate verb forms in the gaps. Use *going to, will, won't, be certain, may, might.*

A

I (1) *'m going to* become a lawyer. I (2)_____ that I've done well in my exams and my father's old firm (3)_____ probably want to interview me. I'm sure I (4)_____ make a good impression and, in fact, they (5)_____ probably offer me the job on the spot. So I have made up my mind – I (6)_____ sell my house in Folkestone and buy a flat in London.

B

Perhaps I (1)_____ *will* _____ apply for a position in a lawyer's office but it's possible I (2)_____ be offered a job. So I (3)_____ try to study law at university. But maybe I (4)_____ fail the exams. I suppose I (5)_____ pass and so I think I (6)_____ write and ask for an application form.

6 Match a sentence from A with one from B.

A

1 That model looks bored. ☐

2 Your mother sounded quite angry. ☐

3 What a strange accent they have! ☐

4 Poor little dog! His ear's bleeding. ☐

5 Is that your brother? He looks like a ☐
   politician.

6 He said 'beat' instead of 'meat' and ☐
   'bad' not 'mad'.

7 The map shows some sort of building ☐
   round here.

8 Do you know this music? It sounds like ☐
   Mozart.

B

a Actually he's a hotel manager.

b How clever of you. You're quite right.

c He sounds as if he's got a cold.

d There's an old church. It looks empty.

e How long has she been sitting there?

f They sound as if they're from Australia.

g He looks as if he's been in a fight.

h I've told her I'm moving into a flat.

7 Read what happened on *A Wedding Day to Remember* and complete the gaps with correct forms of *can, can't, could, couldn't, be able to.*

My dad was driving me to the church on my wedding day when suddenly there was an awful noise – we had had a puncture. We were on a main road five miles out of town. My dad (1)_____ push the car to the side of the road and we looked at each other. I wanted to cry, but with a great effort I (2)_____ stay calm. Everyone else was in the church at this time so we (3)_____ phone anyone to warn them we would be late. '(4)_____ you change a wheel, Dad?' 'Yes, of course, but not in these clothes. You must get to church. You'll have to hitchhike.' I stood by the road

in my long white dress and in a few minutes I got a lift. The driver was very friendly and I invited him to my wedding, but he said, 'Thank you, but I (5)_____ stop. I'm getting married myself this afternoon and I don't want to be late.' Finally, I arrived at the church and walked down the aisle and all the guests (6)_____ see my beautiful dress. I hoped they (7)_____ see any signs of mud on it.

8 Complete the gaps using *look, look like* or *look as if* and *can't* or *must*.

1 I'm sorry I'm late. You __*look*_____ annoyed. You __*must*_____ have been waiting for ages.

2 You _____ tired. You _____ be ready to go to bed.

3 She _____ a schoolgirl, I know, but actually she's much older than she looks. She _____ be about 23.

4 This house _____ empty. They _____ live here after all.

5 He _____ he didn't sleep last night. He _____ be capable of doing a day's work.

6 You _____ freezing. You _____ be wearing enough warm clothes.

7 Here's an interesting picture. It _____ a photograph, but it is really a painting. It _____ be very expensive.

8 The children _____ they haven't been washed for a week. Their parents _____ look after them properly.

WRITING

> MONSTER ENGLISH LANGUAGE PROGRAMME
> Learn English on the banks of Loch Ness!
>
> Spend a week on an intensive course where you listen, speak, read, write and maybe dream only in English. Students will share a room with a student from another country and lessons will take place between 7 am and 9 pm. There is no transport provided until the end of the course. Fast improvement promised.

You have read the advertisement for an English course and decided to spend a week there.
First write down the rules that are handed to students on their arrival at the school.
Then keep a diary while you are there and note down each day what you think about the course and about the other students.

Lessons 21–25

VOCABULARY

1 Answer the questions.

1 Name two animals that are black and white. _____ _____

2 Name an animal with a long nose and one with a long neck. _____ _____

3 Name two birds that catch and eat small live animals. _____ _____

4 Name an animal that lives in the sea. _____

5 Name an animal that gave its name to a car. _____

6 Name a wild animal from the dog family. _____

7 Name a bird that eats dead animals. _____

2 Put the names of the animals into the squares and read down the boxed letters to find another animal.

1 An insect with brightly coloured wings.   1 _ _ _ _ _ | _ | _ _ _

2 A huge grey animal with ivory teeth.   2 _ _ _ _ | _ | _ _ _

3 This animal is striped.   3 _ | _ | _ _ _

4 Children love this animal, which is black and white.   4 _ _ | _ | _ . _

5 This animal has spots.   5 _ _ | _ | _ _ _ _

6 A bird that seems to fall from the sky very   6 _ _ _ | _ | _ _
   quickly and is good at hunting.

7 This type of deer lives in North America.   7 _ _ _ _ | _ |

8 A black and white type of horse.   8 _ _ _ | _ | _

9 Thousands of this type of bird live in cities.   9 _ _ _ _ | _ | _

10 An animal that carries its house on   10 _ _ _ _ _ _ | _ | _
   its back.

The other animal is a

_____

32

3 Match an adjective (1-9) with a noun (a-i).

1 appalling   a science fiction film
2 charming   b documentary
3 fantastic   c sitcom
4 funny   d horror film
5 gripping   e quiz show
6 independent   f love story
7 romantic   g thriller
8 spectacular   h musical
9 popular   i action film

4 Circle the odd-word-out.

1 political   musical   regional   national
2 editor   stammer   poacher   viewer
3 amazing   startling astonishing appalling
4 sarcastic   exotic   precious   fantastic
5 cynical   lethal   enamel   cruel
6 tabloid   satellite magazine   broadsheet

5 Which word in each line has a different vowel sound?

1 cute   music   view   busy
2 moon   rose   loop   prove
3 fierce   grief   thief   leaf
4 trail   whale   male   talent
5 cruel   jewel   who'll   hole
6 alive   forgive impressive

6 Change one word in each sentence so that it makes sense. Choose from these words

viewers broadcast shouting independent
breathed circulation

1 The blind man ~~stared~~ deeply when he discovered the scent of the rose.

   breathed

2 The number of readers of our most popular soap opera has increased.

   _____

3 Newstime is a weekly magazine with a station of twenty-five thousand.

   _____

4 At six o'clock there's a live American chat show published by satellite.

   _____

5 The Radical is a left wing broadsheet without any political bias.

   _____

6 When he heard her whispering for help, the poacher put down his gun and ran to the river.

   _____

7 Choose the correct word.

1 He gave me a bunch of *chrysanthemums/kissograms*.

2 Tigers are in danger of becoming a(n) *distinct/extinct* species.

3 The diamond necklace was *exotic/exquisite*.

4 A *herd/pearl* of elephants moved quickly past me and disappeared into the jungle.

5 'I fell in love with your brother,' she said *faithfully/truthfully*.

8 What am I? Complete the sentences with the following words and answer the question.

archetypal exotic exquisite extinct lethal
loops raise terror waterhole

1 You may meet me at a _____ .

2 I come from the heart of the _____ jungle.

3 I can relax and twist myself into a series of _____ .

4 Or I can _____ my head and look around me.

5 Then I create fear and _____ in all living creatures.

6 I am the _____ symbol of evil.

7 I am _____ when I bite.

8 I am hunted for my skin, which some call _____ .

9 Be careful not to take too much or I will become _____ .

10 The sound I make is like the first letter of my name. What am I?

_____

9 Correct the mistakes. In each of the sentences there is a multi-part verb and one of the parts is wrong. Write the correct verb.

1 The plane ~~took up~~ from the runway.

   took off

2 You can't ignore things any longer. You must face on to your problems and accept help.

_____

3 Do you go on with your sister?

_____

4 The policeman jumped on his bicycle and was able to catch up to the thieves.

_____

5 How long does the smell of burning rubber take to go off?

_____

6 She had a week off work because she fell down with flu.

_____

7 I hope they're going to do after with whale-hunting. It shouldn't be allowed.

_____

8 Her boss tried to give her worse pay than her male colleagues but she stood out to him and insisted that she received the same.

_____

9 When are you going to make up watching this stupid programme?

_____

10 Our plans for our holiday have fallen out. The travel agency has lost all its money.

_____

11 Why don't you go in with the competition for writing poetry?

_____

12 I often have to work late in my job, but the long holidays make up with that.

_____

10 Choose the correct word.

1 Yesterday I noticed a _____ young woman wearing a fur coat going into Mr Jenkins' jeweller's shop.
   a precious   b brilliant   c well-dressed
   d expensive

2 In the window there were some boxes made of ivory and necklaces made of _____ .
   a tears   b pearls   c loops   d antiques

3 Suddenly there was a demonstration in the square and crowds of people were shouting 'Save endangered _____ !'
   a species   b servants   c sciences
   d sanctuaries

4 The woman came out of the shop looking
   _____ .
   a fainting   b wobbling   c dusky   d dazed

5 The crowd shouted at her _____
   and a policeman had to lead her away.
   a sympathetically   b reluctantly   c fiercely
   d finally

6 I was extremely _____ and wished I
   had been brave enough to help.
   a awkward   b startled   c incredible
   d clumsy

7 Then I heard a(n) _____ behind me and
   turned round to see the television cameras.
   a click   b intrigue   c bell   d scent

8 They were filming a scene for my favourite
   _____ showing some of the
   characters taking part in an animal rights
   campaign.
   a sitcom   b edition   c quiz show
   d soap opera

9 The programme will be broadcast next
   year. They warned me I would probably
   appear on _____ .
   a documentary   b TV   c video   d satellite

GRAMMAR

1 Complete the sentences with a suitable
   adverb.

   1 'You're i_____ brave,' she told
     him, as he jumped into a river full of
     crocodiles.
   2 'I have been p_____ in love with
     you from the moment I saw you,' he
     exclaimed.
   3 The acting in the film about King George
     was a_____ brilliant.
   4 I found the ending d_____ moving. I
     was crying when I left the cinema.

5 I laughed out loud at the scene in the
   transport café. It was p_____
   funny.

6 The TV documentary about unemployment
   revealed an e_____ accurate
   picture of the experience of young people
   today.

2 Read the following paragraph. Add these
   words in the appropriate places in the
   story. You will need to change some of the
   adjectives into adverbs.

   bored   calm   clumsy   comfortable   immediate
   loud   nervous   noisy   patient   peaceful
   quick   quiet   reluctant   frightened

   After dinner the whole family went out to sit
   in the garden. The sun was setting and the
   view from the top of the hill made a
   (1)_*peaceful*_ scene. But Clarissa was
   (2)_____ and wanted to go out with
   her friends. Her father said no, so Clarissa sat
   down (3)_____ on the ground and
   said nothing. Her mother was listening
   (4)_____ to her own mother while
   making sure she was (5)_____ settled
   in her chair. The boys were playing cricket
   (6)_____ in front of the house.
   Robert tripped and threw the ball
   (7)_____ over Paul's shoulder. Paul
   ran (8)_____ into the trees to look
   for it. For a short time everything was
   (9)_____ . Then Clarissa heard her
   father saying, 'Don't be (10)_____ but
   I saw something moving behind that tree a
   few seconds ago.' Clarissa looked round
   (11)_____ , then stared in horror. She
   saw two eyes staring back. She screamed
   (12)_____ . (13)_____ the
   eyes disappeared. Her father was standing at
   her side. 'I'd better phone the police,' he
   announced (14)_____ . 'A leopard
   escaped from the local zoo this morning.'

3 Read the following statements made by Dr Peregrine Moorhouse, a famous scientist. Choose an appropriate reporting verb and change the sentences into reported speech.

announce   say   explain   hope   promise

1   Whales communicate with each other by making noises.

He said that whales communicate

with each other by making noises.

2   We have set up an experiment to monitor the behaviour of 500 whales.

_____

_____

_____

3   We are going to start recording their sound next week.

_____

_____

_____

4   Tomorrow I hope to be able to tell you the full amount of money available for this project.

_____

_____

_____

5   The similar experiment that I conducted last year failed because it was the wrong time of year.

_____

_____

_____

6   This time we will not fail.

_____

_____

4 Read the following dialogue and change it into reported speech.

Tom   How many Valentine cards did you receive this year?

Tom asked

_____

_____

Sara   Ten.

_____

Tom   Who were they from?

_____

Sara   I don't know.

_____

Tom   Why didn't you send any?

_____

_____

Sara   How do you know that I didn't send any?

_____

_____

Tom   Which card did you like best?

_____

_____

Sara   The one with the rose on the front.

_____

_____

Tom   Do you know who it's from?

_____

Sara   No, I don't.

_____

Tom   Did you like the one with the heart on the front?

_____

_____

Sara   How do you know I got one with a heart on the front?

_____

_____

Tom   Surely you recognised my handwriting!

_____

_____

5 Read the account of a conversation. Write out the three parts of the conversation in direct speech.

A Simon told me that there was a good job advertised in the paper and advised me to apply. I agreed to think about it but said I wasn't sure it was the right job for me. Simon encouraged me to be more confident and finally persuaded me to send off an application.

Simon _____

_____

_____

You _____

_____

Simon _____

_____

You _____

B A few days later I told Simon that they had invited me to go for an interview. I promised to phone him after the interview. He reminded me to smile and look relaxed and he asked me what I was going to wear. I decided to wear my navy suit and I hoped they wouldn't notice there was a button missing. Simon suggested I should sew one on before the interview.

You _____

_____

_____

Simon _____

_____

You _____

_____

Simon _____

_____

C After the interview I phoned Simon to say it had gone really well and they had offered me the job. I said I would have to move to Glasgow. Simon warned me that I would be lonely at first but promised to come and visit me very soon.

You _____

_____

_____

Simon _____

_____

_____

WRITING

Choose a film that you enjoyed watching. Describe the characters and the plot. Give your opinion of the film and say what your friends thought about it and what the critics said about it in the newspaper.

Lessons 26-30

VOCABULARY

1 Look at the words in the box and match them with the definitions below.

beef grill goulash bake fry potato
pepper hamburger crab roast spaghetti
mussel avocado lobster onion pork corn
tomato boil hot dog oyster

1 Two kinds of meat.

beef _____ p_____

2 Two cooked dishes.

g_____ s_____

3 Two types of fast food.

h_____ h_____

4 Four types of seafood.

c_____ l_____

m_____ o_____

5 Four types of vegetables.

p_____ p_____

o_____ c_____

6 Two types of fruit.

a_____ t_____

7 Name five ways to cook food.

b_____ b_____

f_____ g_____

r_____

2 Your friends are on holiday. They have written a postcard to tell you about it. Use these words to complete the sentences.

afraid beach island swimming pool
thunderstorm wreck

We're having a wonderful time.

Today we climbed a mountain but were caught in a
(1) _____ on the way down.

We walked all the way round the
(2) _____ but we lost our way back
to the hotel.

We can't swim in the hotel (3) _____
because it's empty and we're (4) _____
to go out at night because there have been some
muggings locally.

We can't sunbathe because the (5) _____
is dirty.

We sailed round the island and hit a
(6) _____ under water.

Wish you were here.

Love Pam and Phil.

3 Circle the odd-word-out.

1 tea   coffee   water   wine
2 gold   steel   coal   clay
3 rice   banana   wool   wheat
4 coffee   tobacco   cotton   china
5 milk   cider   beer   wine

4 Are these sentences true (T) or false (F)?

1 Ships are built in Glasgow. ☐
2 Scotch is made in Scotland. ☐
3 China is imported from China. ☐
4 Cider is made from apples. ☐
5 Distilleries are situated in Sheffield. ☐
6 Cutlery was manufactured in Sheffield. ☐
7 Coal was mined in Wales. ☐
8 Tobacco was discovered in the USA. ☐
9 Computers were invented in Japan. ☐
10 Electricity is produced by nuclear power. ☐

5 Correct the mistakes. There is one in each question. Cross out the wrong word and write the correct word.

1 The garden is overgrown. We need a plumber.

_____

2 The path is blocked in the kitchen. We need a plumber.

_____

3 The shower isn't working. It needs overflowing.

_____

4 The wallpaper is dripping. We need a decorator.

_____

5 The wiring's old-fashioned. We need a carpenter.

_____

6 The carpet's scratched. It needs cleaning.

_____

6 Match each sport with a place a-f.

1 athletics   a pool
2 tennis   b ring
3 boxing   c court
4 football   d lane
5 skiing   e field
6 swimming   f slope

7 Which sports are these words connected with?

1 boots

_____

2 racket

_____

3 gloves

_____

4 referee

_____

5 match

_____

8 Put one of the adjectives in each of the following sentences.

dangerous difficult easy essential expensive
important interesting unnecessary impossible

1 It is _____ to score goals.

2 It is _____ to keep your eye on the ball.

3 It is _____ to ski in the summer.

4 It is _____ not to wear head protection.

5 It is _____ to go on a skiing holiday.

6 It is _____ to spot new talent.

7 It is _____ to practise.

8 It is _____ to run a few metres.

9 It is _____ to buy all your own equipment.

9 Read the following instructions and recommendations. What are the people doing?

1 Make sure you get everybody in the picture.

 Taking a photo

2 Always take a taxi home afterwards.

 _____

3 Wear a suit and be punctual.

 _____

4 Always admit to any damage when you give it back.

 _____

5 Don't forget to write and thank everyone for the wedding presents.

 _____

6 Never put jeans and white things together.

 _____

7 Try to keep everything hot.

 _____

8 Don't forget to take one three times a day.

 _____

10 Complete the sentences with forms of *do, give, have, take.*

1 We _____ a party last week. Somebody _____ some damage to the carpet.

2 I'm going to _____ you some advice: _____ a taxi home.

3 After I've _____ the ironing, I'm going to _____ a bath.

4 _____ you time to _____ me some help?

5 _____ your time and _____ your best.

6 I'll _____ her a ring before I _____ the shopping.

7 Don't you want to _____ your hair before I _____ a photo?

8 Will you _____ the washing up while I _____ the beds?

9 I'm trying not to _____ any mistakes because I don't want to _____ the test again.

10 I asked her to _____ me an example and she _____ her best to think of one, but she couldn't.

GRAMMAR

1 Patty Palmer is a famous swimmer. She was interviewed recently for a magazine. Read her comments and fill in the questions. Use the words in brackets.

1 *Who was the person who influenced you most?*

> When I was still at school I was training for the Olympics. The other teachers were angry if I didn't do my homework but Miss Appleton always asked me about my swimming first and said I must choose for myself the most important things in my life. I never forgot that advice.

2 (*person*)_____

> I'd like to meet Nelson Mandela. He's had to endure so much in his life. I really admire him for coping with so much change and continuing to work so hard even when he's quite old.

3 (*event/change*)_____

> One day I went to school as usual and then I went swimming with my friends. I was quite small and nobody noticed me usually. One of the boys asked how fast I could swim one length. We both jumped in and I beat him quite easily. He looked at me in amazement and I knew then what I was going to do with my life.

4 (*book/enjoy*)_____

> I read *A Tale of Two Cities* when I was sixteen and I thought it was the most wonderful book in the world. I cried and cried so it's strange to say I enjoyed it.

5 (*place/grow up*)_____

> I was born in South Africa but spent my childhood in Sheffield, a city in the North of England. You can quickly get to the countryside from there and my friends and I used to go on long bike rides.

6 (*people/dog/rescue*)_____

> They are the kindest people I've ever met. They're called Jean and Frank Swan. Since I spotted their dog in the middle of the river and I swam out to rescue it, they have given me money to buy clothes and equipment and they send me a card before every race.

7 (*expensive*)_____

> I suppose it's my car. I really enjoy driving and when I won my first prize money I bought a new sports car. I love it.

Now answer questions 1–5 and 7 with personal information.

1 _____

_____

2 _____

_____

3 _____

_____

4 _____

_____

5 _____

_____

7 _____

_____

2 Match words from A and B. Make sentences in the passive by adding verbs and prepositions.

A	B
sushi	Marconi
cricket	millions of viewers
radio	thousands of years
whisky	India
Diana's funeral	Japan
languages	next year
rockets	Scotland
their wedding	from Cape Canaveral
	since 1961

1 *Sushi is eaten in Japan.*

2 _____

3 _____

4 _____

5 _____
_____

6 _____
_____

7 _____
_____

8 _____
_____

3 You are preparing dinner for ten people. You are late and you still haven't done a lot of things. Somebody offers to help you. Suggest some things they could do.

1 peel the vegetables

The vegetables need peeling.

2 wash the fruit
_____

3 chop the meat into small pieces
_____

4 put the pie into the oven
_____

5 put some cutlery on the table
_____

6 pour the wine into the glasses
_____

4 You are getting married soon. Your mother is arranging everything. Remind her to do things. Use these verbs.

send out   order   arrive   make   clean   order   book   send

1 (don't forget/flowers)

Don't forget to order the flowers.

2 (make sure/church)

Make sure you arrive at the church on time.

3 (make sure/invitations)
_____
_____

4 (don't forget/guests/a map)
_____
_____

5 (don't forget/the bridesmaid's dresses)
_____
_____

6 (make sure/my father's suit)
_____
_____

7 (don't forget/a cake)

_____

_____

8 (make sure/the reception)

_____

_____

5 You have bought a house. Look at the list of things that need doing and look at your diary. Make sentences about next week.

MONDAY

builder

_____

TUESDAY

plumber

_____

WEDNESDAY

decorator

_____

THURSDAY

carpenter

_____

FRIDAY

electrician

_____

SATURDAY

gardener

_____

SUNDAY

I'm busy in my new

light-switch — mend
grass — cut
bath — replace
whole house — clean
walls — replaster
door — fix
bedrooms — paint

1 *I'm having the walls replastered on Monday.*

2 _____

_____

3 _____

_____

4 _____

_____

5 _____

_____

6 _____

_____

7 _____

_____

6 Answer the questions using reflexive pronouns.

1 Do you make your own bed?

Yes, *I make it myself.*

2 Do you and your wife get a gardener to do your gardening?

No, _____

3 Did they have their sitting-room decorated?

No, _____

4 Did you help her fix the lawnmower?

No, _____

5 Has he got a dishwasher?

No, _____

6 Can I have this letter typed?

No, _____

7 Janet is staying at a health clinic. She is not enjoying herself! Katie is on holiday on a farm. She loves animals. Write what they say using *make* or *let* and these phrases.

get up at 6 am	help with the chickens	take cold showers
ride the horses	do exercises	feed the pigs
smoke	stay up late	milk the cows

JANET

1 *They make me get up at 6 am.*

2 _____

3 _____

4 _____

5 _____

KATIE

1 *They let me ride the horses.*

2 _____

3 _____

4 _____

WRITING

Imagine you have been to look at the house in this advertisement. Is it your dream house? Write a letter to a friend explaining why you are or why you are not going to buy it.

> HOUSE IN THE COUNTRY
>
> This is a delightful five-bedroomed house about 30 minutes on foot from the railway station. There is no road and it is completely private, with fine views of the surrounding mountains. The whole house is newly decorated; there is central heating and modern wiring in all rooms. There is a swimming pool in the garden but it requires some repairs.

Lessons 31–35

VOCABULARY

1 What do you need to take with you? Write down three things that you need in each situation. The first one has been done for you.

1 You are going to a remote spot in the country to go birdwatching. You may feel hungry.

   a map, some binoculars and

   some chocolate

2 You are going on a train journey and you need something to read which may make you cry.

   a t_ _ _ _ _, a p_ _ _ _ _ _ _ n_ _ _ _

   and a h_ _ _ _ _ _ _ _ _ _

3 You are going out in your best clothes to meet the man/woman of your dreams, who may want to see you again.

   a m_ _ _ _ _, a d_ _ _ _ and a

   b_ _ _ _ _ _ _ p_ _

4 You are travelling up to Scotland overnight on the sleeper train, to go to a meeting, and then to a business lunch.

   an a_ _ _ _ c_ _ _ _, a b_ _ _ _ _ _ _

   c_ _ _ and a c_ _ _ _ _ c_ _ _

5 You are going to an outdoor pop festival. It is raining. You will be able to buy bread, cheese and fruit.

   an u_ _ _ _ _ _ _ , some c_ _ _ and a

   p_ _ _ _ _ _

6 You provide a shopping service for some of your friends. You go to a big shopping centre with their shopping lists, you buy the things and then deliver them to their homes.

   a n_ _ _ _ _ _ _ , a c_ _ _ _ _ b_ _ _ and

   an a_ _ _ _ _ _ b_ _ _

7 You are travelling home by train and you are late. You will have to drive straight from the station to the concert hall for the opera. You have not got time to pick up your partner first.

   your car k_ _ _ , two o_ _ _ _ t_ _ _ _ _ _

   and a m_ _ _ _ _ p_ _ _ _

2 Rearrange the letters in the following words and use them to answer the questions.

knec liwlop ecnsit lelenetrp ordo gewed
iglth blub cohtr reaslpgu

1 What do you need to replace if the room is too dark? _____

2 What do you need if you camp in a hot, damp climate? _____

3 What can help you sleep on a bus or a train? _____

4 What can you use to hold a door open or closed? _____

5 What can help you sleep in a noisy room?

   _____

6 You are trapped in the dark without one.

   _____

3 Correct the mistakes. One word is wrong in each of the following sentences.

1 Lift the boot when you want to look at the engine.

_____

2 Use the wipers to keep the dashboard clear when it's raining.

_____

3 The indicator tells you how fast you are going.

_____

4 Check the exhaust to see if you have enough petrol.

_____

5 There's plenty of room in the bonnet for your luggage.

_____

6 You use your left foot only on the accelerator.

_____

7 Always carry a steering wheel in case you have a puncture.

_____

8 You'll need a pedal if you're going to change a wheel.

_____

9 My driving instructor says you shouldn't drive with your hand on the rear light.

_____

10 Oops! You've reversed too far and I think you've hit his brake.

_____

4 Choose the correct word in each sentence.

1 Use dry white wine to remove a _____ .
   a fly b aroma c stain d rag

2 Could I have two _____ of sugar in my coffee?
   a marks b lumps c pieces d limps

3 _____ the chewing-gum after putting the item in the freezer.
   a Stretch b Mop up c Brush off d Scrape off

4 The dog settled down in her basket with her six new _____ .
   a puppies b blankets c bottles d torches

5 The best way to treat hiccups is to _____ water while you hold your nose.
   a soak b sip c pour d drip

6 If you are trying to sell your house, make sure visitors can smell the _____ of coffee.
   a ash b ring c lid d aroma

5 Choose the best definition for each word.

1 **equator**
   a something that is equal in value to something else
   b an imaginary line around the centre of the world
   c a person who travels across the sea

2 **propaganda**
   a information given to the public to make them believe something
   b the correct way of behaving in a social group
   c a type of Chinese house

3 **solo**
   a close to the ground
   b in a male-dominated sport
   c alone; by one person on their own

4 **drown**
   a to die because water prevents air reaching the lungs
   b to drop out of the air before flying up again
   c to disappear

5 **rumours**
   a items of clothing that form part of a uniform
   b stories that may or may not be true
   c stories about travelling

6 **trace**
   a a message
   b a map
   c a sign that somebody or something exists.

6 Read the following newspaper article. Complete the gaps using these words.

appreciate avoid caused cheap cultural economic embarrassed encourage environmental hospitable out of season experience public sights

The small town of Popplethwaite is suffering serious (1)_____ effects from the decision made by the government not to build a second Channel Tunnel nearby. Its inhabitants had been promised very little (2)_____ damage and huge financial rewards. Many firms have now cancelled their plans to open new businesses in Popplethwaite. The only hope now is if all local people join together to (3)_____ tourism. Visitors must be allowed to (4)_____ the beautiful countryside and to get to know the (5)_____ background of the town.

There is (6)_____ accommodation available in the hostels prepared for the tunnel workers and tourists will be made welcome in the shops, pubs and restaurants run by friendly, (7)_____ people. There is also an efficient (8)_____ transport system and a number of historical (9)_____ and monuments as well as huge beaches, crowded in summer but often empty (10)_____ .

Last year a group from Popplethwaite visited a town in Belgium with the aim of establishing a link. Unfortunately the host families complained that their guests made no efforts to (11)_____ excessive consumption of the local beer. Several people (12)_____ their hosts by insisting on making tea with their own teabags. The whole incident (13)_____ some resentment among the Belgians who thought the aim of the visit was for the British to (14)_____ the local lifestyle.

7 Look at the following words. Find pairs of words and put them in the correct column.

disappearance distant protection survival exhaustion mysterious distance arrive frustration feminist speculate lecture mystery prominent disappear correspondent lecturer indifference exhaust famous correspondence feminism protect fame frustrate survive indifferent speculation prominence arrival

VERB - NOUN

disappear - disappearance

_____

_____

_____

_____

_____

_____

NOUN - ADJECTIVE

distance - distant

_____

_____

_____

_____

NOUN - NOUN

_____

_____

_____

GRAMMAR

1 Make sentences using *if* or *in case*.

1 take / some sandwiches / be / hungry
   before you get there

   Take some sandwiches in case

   you're hungry before you get there.

2 fuel gauge / on the red mark / stop / at
   the next petrol station

   _____

   _____

3 the steering wheel / look / well-used / I /
   always / check the mileage

   _____

   _____

4 join / a motoring organisation / your car /
   break down / a long way from home

   _____

   _____

5 I / switch on / the fog lights / be / foggy

   _____

   _____

6 A light on the dashboard / come on / you
   / forget to fasten your seat belt

   _____

   _____

7 I / take / my cheque book / the petrol
   stations / not take credit cards

   _____

   _____

2 How do you do it? Rewrite these sentences
using the word in brackets.

1 To go faster press the accelerator. (if)

   If you press the accelerator you

   can go faster.

2 You can avoid the motorways if you follow
   a map. (by)

   _____

   _____

3 By using a neck pillow you can easily
   sleep on trains and planes. (if)

   _____

   _____

4 If you use insect repellent you can protect
   yourself from mosquito bites. (by)

   _____

   _____

5 To stop hiccups hold your breath and
   count to ten slowly. (by)

   _____

   _____

6 By spinning an egg you can tell if it is
   hard-boiled. (to)

   _____

   _____

3 Match a sentence from A with one from B. Then rewrite the sentences using the words in brackets. Look at the example first.

They told me I could collect the car on Friday. (due to)

First I'll drive home and then we can all go to the seaside. (as soon as)

I'm due to collect the car on Friday. As soon as I've driven home, we can all go

to the seaside.

A

Their train gets in at half past seven. (due to)

We estimate the time taken to reach the start of the river will be five days. (likely to)

His teachers think he will pass all his exams. (expected)

You will probably hear strange noises in the night and you must keep absolutely still. (when)

The doctor will examine him and then she will tell us what is wrong with him. (after)

Don't leave the room until she falls asleep. (as soon as)

B

We can't confirm his appointment until we hear his results. (when)

I think he will stay in hospital for at least a week. (likely to)

They are going to phone me from the station and then I will start cooking. (as soon as)

My husband and I plan to return at about midnight. (due to)

We will set up camp and then we can explore the jungle. (after)

Any sound will probably be made by a bear. (likely to)

1 _____

_____

2 _____

_____

3 _____

_____

4 _____

_____

5 _____

_____

6 _____

_____

4 Complete the sentences in the first conditional.

1 leave / your phone number / ring you / when it's ready

_____

_____

2 she / grow / any taller / be / the tallest in the family

_____

_____

3 not / pass / this time / I / give up / trying

_____

_____

4 he / continue / driving at this speed / beat / the record

_____

_____

5 you / give / me the recipe / I / make it at home

_____

_____

6 you / embarrass / your hosts / they / not invite / you again

_____

_____

7 taxi / not come soon / I / miss my flight

_____

_____

5 Put the verbs into the correct tense.

1 After she (pour) white wine on the stain, she (wait) ten minutes.

After she had poured white wine

on the stain, she waited ten minutes.

2 They (eat) as much as they (can), and so they (call) the waiter to ask for the bill.

_____

_____

3 She (tell) me she (see) a ghost.

_____

_____

4 He (fall) into the water and (drown).

_____

_____

5 The policeman (tell) her she (speed) .

_____

_____

6 When we (travel) for a week, I (begin) to feel sick.

_____

_____

7 He (wait) since January and it (be) now October.

_____

_____

8 She (not stop) polishing until she (make) everything shiny.

_____

_____

9   The spy (recognise) at the airport because
    he (forget) to put on his glasses.

10 She (decide) to join a feminist movement
    because she (feel) she (always live) in a
    male-dominated world.

_____

_____

_____

_____

WRITING

Last year you spent three months living in the Andes. Describe your experiences and advise a
friend how to prepare for a similar trip. Use the following words: *footprint, strange light,
camera, binoculars, paperback novel, phone home.*

Lessons 36–40

VOCABULARY

1 Put the names of the schools in the correct order in both Britain and the USA.

junior high school   middle school   university
nursery school   elementary school
comprehensive school   university
primary school   kindergarten   high school

GB *nursery school* _____
_____
_____
_____

US _____
_____
_____
_____

2 What are the university subjects?
Rearrange the letters to find out.

1 shytroi _____

2 hyiscps _____

3 wal _____

4 gghrpaoye _____

5 syctiermh _____

6 ahsmt _____

7 deecnimi _____

8 ococsenim _____

9 sohopyliph _____

10 esnagalug _____

3 Circle the odd-word-out.

1	summer	autumn	winter	season
2	anchor	mall	float	moor
3	bikini	miniskirt	junk	blouse
4	bamboo	silk	porcelain	lobby
5	stride	frown	yawn	chatter
6	curse	threaten	beckon	routine

4 Choose the correct definition.

1 **scarce**
   a in high demand
   b in short supply
   c easily frightened

2 **dirt-cheap**
   a very dirty
   b very good value
   c costing very little money

3 **shortage**
   a too little time
   b an inadequate supply of something
   c the fashion of wearing miniskirts

4 **manhole**
   a a hole in the road or pavement that gives access to workmen
   b a hole in the ground that is as big as a person
   c that can be operated without machinery

5 **reincarnation**
   a repeating something someone else said
   b the idea that the soul of a dead person enters another body and is born again
   c aliens from another planet visiting earth

6 **cemetery**
   a a place where dead people are buried
   b a large church
   c an artist's studio

5 Put an appropriate form of one of these idioms into each of the following sentences.

cannot make head nor tail of
be all ears
be all fingers and thumbs
put your foot in it
put your finger on something

1 I knew as soon as I walked into the room that I'd been there before. I _____ _____ but the whole place was familiar.

2 I can't help you with your model aeroplane, I'm afraid. I _____

3 The letter's in Italian and I can't read her handwriting so I _____ _____

4 Oh no! I _____ . I told her I thought she was older than her sister.

5 Go on, tell me about your first bungee jump. I _____

6 Replace a word or words in each sentence with the word in brackets. Circle the word(s) you would replace.

1 His first (meeting) with the philosopher happened quite by chance. (encounter)

2 She remained completely unaware of his romantic feelings about her. (oblivious)

3 I can't stand the way that fellow is looking at you. Isn't he one of your colleagues? (guy)

4 It's best if you pay no attention to the camera. I will film the whole event and select the best bits later. (ignore)

5 Make sure you put a cover on the saucepan before turning down the heat. (lid)

6 My friend has kindly agreed to come to the ceremony with me and help me with interpretation. (companion)

7 She had noticed some very strange behaviour in him recently. He wore casual shoes to work and carried an ancient umbrella even when it wasn't raining. (odd)

7 Use these words as often as necessary to fill the gaps in the following sentences. Then put the sentences in the correct order to complete this letter to a problem page.

ask   break   fall   get   go   engaged involved
married   in   out   jealous

☐ a I realised I fancied him again when I started *feeling* _____ .

☐ b He _____ me _____ in the third year of comprehensive school.

☐ c 'I am _____ love with Jenny!' he explained to me.

☑1☑ d I _____ to know Simon when we were at school together.

☐ e We _____ up and didn't speak to each other for five years.

☐ f How can I let him _____ _____ to her? Please help me.

☐ g We _____ _____ together for two years.

☐ h I started seeing him behind her back but he said he didn't want to _____ _____ with me.

☐ i I _____ _____ with him when I saw him dancing closely with my best friend Anna at a party.

☐ j Then we met again by chance and he told me he had _____ _____ to my best friend Jenny.

8 Complete the gaps in the following sentences.

1 Look at him lying across the table. He's as drunk as a _lord_ .

2 She grew up near the sea. Since she was three she's been able to swim like a _____ .

3 Why don't you listen to what I'm saying? You really are as _____ as a mule.

4 He's been _____ like a dog all this week so he can go on holiday next week.

5 They've eaten all the chocolate cake! They're as _____ as pigs.

6 I'm going to sleep like a _____ tonight after such a long walk in the mountains.

7 She didn't pass a single exam. She must be as _____ as two short planks.

8 You starve all day and then you eat like a _____ in the evening. You'll never lose weight.

GRAMMAR

1 Make sentences in the second conditional.

1 I / see / him again / ask / him his name

If I saw him again, I would ask him his name.

2 you / go / to Hong Kong / see some spectacular scenery

_____

_____

3 you / wear / flip-flops to school / the headmaster / send / you home

_____

_____

4 you respect / your best friend's feelings / not flirt / with her boyfriend

_____

_____

5 supply / exceed / demand / the company / lose / a lot of money

_____

_____

6 you / pass / your exams / be able to / go to university

_____

_____

7 I / see / a ghost / think / I was going crazy

_____

_____

2 Give advice to a friend in reply to the following comments.

1 'The tiles round the bath are faded and cracked.'

If I were you _____

_____

2 'I keep hearing strange noises in my head.'

I think _____

_____

3 'I received a mysterious letter threatening me with disasters if I don't reply.'

If I were you _____

_____

4 'Someone I fancy has asked me to go out with them.'

I think _____

_____

5 'There were three murders in our street last week.'

If I were you _____

_____

3 Reply to the following remarks using the word in brackets.

1 'I've just climbed six flights of stairs and I'm exhausted.' (elevator)

You should have used the elevator.

2 'All the hotels were full and we couldn't find any accommodation.' (peak season)

_____

3 'It's too late to add this document to the letter.' (seal)

_____

4 'We came straight off the plane and were first in the queue for a taxi.' (trouble/customs)

_____

_____

5 'I wish I could go swimming.' (bikini)

_____

6 'There were ten of us in a small compartment with all our luggage.' (cramped)

_____

7 'He didn't hear what I said.' (hearing-aid)

_____

_____

8 'My feet got extremely uncomfortable.' (high heels)

_____

_____

9 'I didn't understand a word that they were saying.' (English)

_____

_____

4 Answer the following questions using the words in brackets and *may have, might have, could have* in turn.

1 Where are Ben and Jemima? (go sightseeing)

_____

2 Where did they meet each other? (in the drugstore)

_____

3 Where have they disappeared to? (down a side street)

_____

4 Where are my old flip-flops? (throw away)

_____

5 Why did she run away?
(feel threatened)

_____

6 How did they get here so quickly?
(get a taxi)

_____

5 Choose five words or phrases and use them to make sentences about how you wish you were different.

ski	in the city
drive	in the country
a bigger house	study economics
a lot of money	like jazz
a film star	opera music
beautiful	good looking

1 _____

_____

2 _____

_____

3 _____

_____

4 _____

_____

5 _____

_____

6 Make sentences in the third conditional.

1 I / study / medicine/ become / a doctor

If I had studied medicine, I would

have become a doctor.

2 they / use / corporal punishment / at my school / I / beat / quite often

_____

_____

_____

3 they / hurry / not miss / the train

_____

_____

4 you / ask out / my best friend / she / accept

_____

_____

5 you / not lose / your umbrella / it / keep / us dry

_____

_____

6 you / tell / me / I / not believe / you

_____

_____

7 you / ask / him for his telephone number / he / think / you fancied him

_____

_____

8 I / have / enough money / buy / a silk dress in Hong Kong

_____

_____

_____

7 Choose a phrase from this list to complete the sentences below. Put the verb in the correct tense.

(call) for help
(have) some beautiful flowers in spring
(not lose) my temper
(not see) a giant tortoise
(give up) my job
(be) in London by midnight
(like to be) a pilot
(have) a dog to keep me company
(lose) my job
(get) sunburn

1 If I won the lottery, *I would give up up my job.*

2 If I had felt in danger, _____

_____

3 Sometimes I get lonely. I wish _____

_____

4 If I am late to work again, _____

_____

5 I was angry, but I wish _____

_____

6 If I left immediately, _____

_____

7 If I lie in the sun too long, _____

_____

8 If I plant these bulbs now, _____

_____

9 If I hadn't become a lawyer, _____

_____

10 If I hadn't gone to the Galapagos, _____

_____

8 Look at these two pairs of sentences. They are similar in meaning.

A If only I had followed him. Then I would have found out where he lived.

B I wish I had followed him. If I had followed him, I would have found out where he lived.

Think about your own life and write two pairs of sentences like A and two pairs like B. Say what you think about the past.

1 _____

_____

_____

2 _____

_____

_____

3 _____

_____

_____

4 _____

_____

_____

WRITING

Look again at the problem page letter in Vocabulary Exercise 7. Write a suitable reply.

Answer key

Lessons 1–5

VOCABULARY

1 1 What is your surname?
   2 How do you spell that?
   3 How old are you?
   4 Are you married?
   5 What's your address?
   6 What's your nationality?/What nationality are you?
   7 What's your favourite sport?
   8 Where do you work?

2 1 d 2 e 3 i 4 a 5 f 6 j 7 c 8 b 9 h 10 g

3 1 preparations 2 engagement 3 traditional 4 bride;
   bridesmaids 5 reception 6 guests 7 wedding
   8 honeymoon

4 1 traffic 2 connection 3 announcement 4 platform
   5 reservation 6 crowds 7 suitcases 8 booking office
   9 taxi rank 10 catch 11 fare 12 round trip

5
Sport	Music	Hobbies
cricket	opera	DIY
tennis	folk	reading
running	lyrics	train spotting
swimming	rehearsal	shopping
walking	classical	bird-watching

6 2 a 3 e 4 g 5 f 6 b 7 d

7 1 boring 2 great 3 all right 4 terrific 5 dreadful 6 exciting

8 2 boring 3 depressed 4 excited 5 frightening
   6 relaxing 7 annoyed/angry 8 shocked/sorry
   9 disappointing 10 thrilled

9 1 c 2 a 3 d 4 d 5 b 6 c 7 a 8 c 9 a 10 d

GRAMMAR

1 1 plays 2 collect 3 are shouting 4 have 5 am meeting
   6 is working 7 wears 8 believe

2 1 Does your mother live with her sisters?
   2 How often do you see your brother?
   3 Which platform does the 6.15 train to London leave from?
   4 What did you give to the bride and groom?
   5 Who is playing the guitar in this rock band?

3 1 doesn't she 2 does he 3 aren't they 4 do you
   5 hasn't he 6 don't they 7 do we

5 2 They have weeks of rehearsals before they perform/before
   performing the music in front of hundreds of people.
   3 We were engaged for six months before we got/before
   getting married.
   4 After they were married/After being married for twenty
   years they suddenly got divorced.
   5 He does odd jobs in the house during his holidays.
   6 The telephone rang three times during dinner.
   7 He plays tennis for three hours every Sunday.
   8 She watched television for an hour/She put television on
   for an hour before she went/before going to bed.

6 1 Could I borrow/Could you lend me a pen, please?
   2 Could you repeat the question?
   3 How do you spell it/that?
   4 What does that word mean?
   5 Could you speak more slowly, please?
   6 How do you say ...... in English?
   7 Would you mind telling me/Would you tell me/Could you
   tell me when we arrive at Sheffield?

7 2 Gunther can't stand playing cricket, but he doesn't mind
   playing rugby.

3 Claude hates decorating, but he likes going bird-watching.
4 Diana can't stand watching TV, but she adores going
   window shopping.
5 Eva detests playing cards, but she doesn't mind making
   cakes.
6 Ingrid likes entertaining and she adores going to
   nightclubs.

Lessons 6–10

VOCABULARY

1 1 spices 2 neighbour 3 banana 4 moody 5 healthy

2 1 confused 2 hospitable 3 foreigners 4 alone
   5 colleague 6 soldiers 7 wallet

3 2 freezing – boiling 3 sweet – savoury
   4 dishonest – honest 5 even – odd 6 grown-up – child
   7 miserable – happy 8 stale – fresh 9 healthy — ill
   10 boring – interesting

4 uncertain, unclear, incomplete, incorrect, inexpensive,
   unfriendly, unhappy, unimportant, unkind, impatient,
   imperfect, impossible

5 1 present 2 outgoing 3 pleased 4 chocolate
   5 optimistic 6 ridiculous 7 neighbour: POPCORN

6 2 d 3 a 4 f 5 g 6 b 7 c

7 2 afford 3 landmark 4 greengrocer's 5 panic 6 invest
   7 cope

8 1 huge 2 hilarious 3 freezing 4 interesting 5 good
   6 miserable 7 boiling 8 angry 9 silly 10 delighted

9 1 week 2 plain 3 hungry 4 Prices 5 waist 6 beach

GRAMMAR

1 1 was carrying 2 started 3 was opening 4 heard
   5 rushed 6 opened 7 switched 8 was walking
   9 smelled 10 ran 11 was pulling 12 caught
   13 was standing 14 carried 15 walked 16 was lying
   17 was eating 18 rang 19 phoned 20 were you doing

2 1 Television programmes used to be in black and white.
   2 There used to be only one channel./There only used to
   be …
   3 People would invite their neighbours round to watch a
   special programme.
   4 There didn't use to be a law about drinking and driving.
   5 You used to go through passport control between England
   and Scotland. (False)
   6 Children would walk several miles to school each day.
   7 There used to be twenty shillings in every pound sterling.

3 1 As soon as/When Ben's mother announced the start of the
   treasure hunt, the children ran into the garden.
   2 When I noticed my friend on the other side of the street, I
   waved.
   3 Just as she swallowed her seventh chocolate, she read the
   number of calories on the side of the box.
   4 When/Just as the plane was landing, she recognised the
   man sitting across the aisle.
   5 As/While he was walking along the road, he whistled a
   tune to himself softly.
   6 As/While I was waiting in the queue for tickets, someone
   tapped me on the shoulder.
   7 As/When they walked off the boat onto dry land, they
   were still feeling sick.
   8 I searched the whole beach until I finally found Jo playing
   with some shells near the water.
   9 She was training to be a teacher until/when a film director
   spotted her.

10 He was looking puzzled until I reminded him that we had met at a party.

4 2 Spices, which improve the taste of many dishes, grow mainly in Eastern countries.

3 Bilbao, where they built a huge new museum, is on the north coast of Spain.

4 Jane Austen, who had a sister called Cassandra, wrote stories about life in quiet country villages.

5 If you like old postcards, go to one of the junk shops, which you can find in every town in Britain.

6 My uncle, who has a sense of humour, sometimes phones me at home and tells me he's a policeman.

7 Edward Whymper, an English explorer, was the first man to climb mountains in Ecuador, where you can still find streets named after him.

5 2 Thank you for bringing me a cup of tea in bed.

3 Please pass me/Would you pass me the salt and pepper.(?)

4 They offered me the job!

5 Didn't I lend you £5 last week?

6 Why don't you take her some fruit?

7 Please send me/Could you send me a photo of your new baby.(?)

8 You didn't leave me your number.

9 A week in Majorca cost us £300.

10 This is Tina White, who is going to read you/us some poems/some of her work.

WRITING

1 *Suggested answer*

Waiter: I'm sorry, sir, but this is a no-smoking area. Would you mind moving to another table?

Customer: I'm terribly sorry. I didn't read the sign.

Waiter: That's all right, sir.

Customer: Excuse me, could you bring me the bill?

Waiter: Here you are, sir.

Customer: I'm afraid you've made a mistake.

Waiter: I'm very sorry, sir.

Customer: Don't worry about it.

Lessons 11–15

VOCABULARY

1 2 Have you paid it/the bill yet?

3 Have you replied to them/the letters yet?

4 Have you picked up the children yet?

5 Have you collected it/the car yet?

6 Have you mended it/the TV yet?

7 Have you wrapped the parcel/posted them yet?

8 Have you thrown them/the newspapers away yet?

9 Have you done the shopping yet?

2 1 evil 2 popular 3 fictional 4 criminal 5 typical 6 success 7 frustration

3 1 rig 2 watch 3 hall 4 control 5 gallery 6 fountain 7 basket

coffee house   cartoon strip   engagement ring prime minister   alarm clock   photo frame   mountain bike

4 1 crew 2 border 3 weapon 4 firefighters 5 currency 6 kidnap

5 2 museum – market 3 skyscraper – cathedral 4 park – square 5 industry – architecture 6 cemetery – population 7 transport – entertainment 8 castle – river

6 1 bored with 2 capable of 3 allergic to 4 pleased with 5 proud of 6 similar to 7 afraid of 8 fond of

7 1 a beautiful antique mahogany table 2 a heavy old Italian violin 3 a brand-new Cartier diamond necklace 4 a large green Victorian sofa 5 a lovely blue porcelain vase

8 1 glad 2 alive 3 alone 4 ill 5 ready 6 sorry

9 1 c 2 b 3 b 4 d 5 b 6 c 7 b

GRAMMAR

1 1 my, yours 2 Hers, his 3 mine 4 their, ours 5 theirs 6 his, hers

2 3 She's already put petrol in the car.

4 She hasn't turned off the electricity yet.

5 She's already watered the plants.

6 She's already bought some maps.

7 She hasn't tidied the house yet.

8 She's already phoned her parents.

9 She hasn't taken out the rubbish yet.

10 She hasn't bought any suncream yet.

3 2 She's been playing the violin for five years.

3 He's been driving since he was 18/for 15 years.

4 She's been talking to her boyfriend/They've been talking since 7 o'clock.

5 I've been feeling miserable since I woke up this morning.

4 2 I've got a pain in my shoulder.

3 I've been playing football.

4 They've been running for the bus.

5 I've been walking in the rain.

6 I've been living in Germany.

5 A Hello, Jan. I haven't seen you at the line-dancing classes for a few weeks.

B Hello, Linda. I haven't had enough time, I'm afraid.

A What have you been doing?

B I've been studying for my English Literature exam.

A Oh really? How many books have you studied/are you studying/have you been studying?

B I've already read three novels by Jane Austen and I'm reading/I've been reading Shakespeare's *King Lear*, but I'm finding it quite difficult.

A Have you read Hamlet?

B No, but I've seen the film.

A Have you read/Have you been reading any poetry?

B Yes, in fact, I've been trying/I'm trying to learn some poems by heart. I've been saying them to myself in the bus on the way to work.

A I like writing poems. I've been writing them since I left school, but I haven't finished one yet.

6 1 English is easier than Chinese.

2 There is less pollution in Rome than in Athens./Rome has less pollution than Athens.

3 *Jurassic Park* is more frightening than *ET*.

4 Liverpool is not as picturesque as Oxford.

5 He has more money than your father has.

6 No, London hasn't got as many skyscrapers as New York.

Lessons 16–20

1 1 white 2 cream 3 beige 4 crimson 5 red 6 navy 7 blue 8 olive 9 green 10 khaki 11 canary 12 yellow 13 orange 14 grey 15 chocolate 16 brown 17 stone 18 black SCARLET TURQUOISE

2 1 ponytail 2 crew cut 3 stubble 4 earrings 5 boxer shorts 6 flip-flops 7 high heels

3 1 A solider, a sailor, a nurse, a police officer, a park keeper, a waitress

2 A gardener, a farmer, a park keeper, a vet

3 A doctor, a nurse, a hospital receptionist

4 A photographer, a designer, an architect

5 A secretary, an office receptionist, a journalist, an engineer, an accountant, a businessman, etc.

4 1 c 2 b 3 d 4 a 5 d 6 b

5 1 with – by 2 make – do 3 nibbled – sucked 4 on – off 5 jacket – pocket 6 heel – toe 7 by – on

6 cool – emotional, fashionable – old-fashioned, flamboyant – reserved, fun – serious, honest – deceitful, lazy – hard-working, prejudiced – fair-minded, sensitive – thick-skinned, shy – self-assured, sophisticated – naïve, young – middle-aged

7 1 cool-headed 2 easy-going 3 world-famous 4 well-behaved 5 strong-willed 6 soft-hearted 7 short-sighted 8 good-humoured 9 left-handed 10 short-tempered

8 2 fill it in 3 turned it down 4 put off 5 pick it up 6 brought him up 7 take off 8 join in 9 take off 10 give up

GRAMMAR

1 1 must 2 should 3 are supposed to, should 4 mustn't, are not allowed to 5 shouldn't, mustn't, are not allowed to 6 are not allowed to

2 2 You are not allowed to smoke.

3 You must not drive faster than/at more than 30 mph.

4 You have to wear a helmet.

5 Visitors are not allowed after 10 pm.

6 You should dry-clean this garment.

7 You must drive slowly because children play here.

8 You are not supposed to take a camera into the auditorium.

3 2 This might be in a restaurant or on a train.

3 This must be on a road in a town.

4 This could be at a cycle race.

5 This may be in a hospital.

6 This must be on an item of clothing.

7 This might be near a park.

8 This could be in a theatre.

4 1 You are not allowed to touch, madam.

2 You should leave bags in the cloakroom, sir.

3 You are not supposed/allowed to bring food and drink into the museum.

4 You aren't allowed to use mobile phones, madam.

5 You are supposed to pay an entrance fee.

6 You should talk more quietly.

5 **A** 2 I'm certain 3 will 4 will 5 will 6 I'm going to
   B 2 won't 3 may, might 4 will 5 might 6 will

6 1 e 2 h 3 f 4 g 5 a 6 c 7 d 8 b

7 1 was able to 2 was able to 3 couldn't 4 Can 5 can't 6 could 7 couldn't

8 2 look; must 3 looks like; must 4 looks; can't 5 looks as if; can't 6 look; can't 7 looks like; must 8 look as if; can't

Lessons 21–25

VOCABULARY

1 1 panda, zebra 2 elephant, giraffe 3 eagle, falcon 4 whale 5 jaguar 6 wolf 7 vulture

2 1 butterfly 2 elephant 3 tiger 4 panda 5 leopard 6 falcon 7 moose 8 zebra 9 pigeon 10 tortoise RHINOCEROS

3 2 h 3 a 4 c 5 g 6 b 7 f 8 i 9 e

4 1 musical 2 stammer 3 appalling 4 sarcastic 5 enamel

5 1 busy 2 rose 3 fierce 4 talent 5 hole 6 alive

6 2 readers – viewers 3 station – circulation 4 published – broadcast 5 left wing – independent 6 whispering – shouting

7 1 chrysanthemums 2 extinct 3 exquisite 4 herd 5 truthfully

8 1 waterhole 2 exotic 3 loops 4 raise 5 terror 6 archetypal 7 lethal 8 exquisite 9 extinct 10 A snake.

9 2 face on to – face up to 3 go on with – get on with 4 catch up to – catch up with 5 go off – wear off, go away 6 fell down with – went down with 7 do after with – do away with 8 stood out to – stood up to 9 make up – give up 10 fallen out – fallen through 11 go in with – go in for 12 make up with – make up for

10 1 c 2 b 3 a 4 d 5 c 6 b 7 a 8 d 9 b

GRAMMAR

1 1 incredibly 2 passionately 3 absolutely 4 deeply 5 particularly 6 extraordinarily, extremely

2 2 bored 3 reluctantly 4 patiently 5 comfortably 6 noisily 7 clumsily 8 quickly 9 quiet 10 frightened 11 nervously 12 loudly 13 Immediately 14 calmly

3 1 He said that whales communicated with each other by making noises.

2 He announced that they had set up an experiment to monitor the behaviour of 500 whales.

3 He said that they were going to start recording their sound the following week.

4 He hoped to be able to tell them the next day the full amount of money available for the project.

5 He explained that the similar experiment he had conducted the previous year had failed because it had been the wrong time of year.

6 He promised that they would not fail this time.

4 Tom asked how many Valentine cards Sara had received this year.

Sara said she had received ten.

Tom asked who they were from.

Sara said she didn't know.

Tom asked why Sara hadn't sent any (cards).

Sara asked how Tom knew that she hadn't sent any (cards).

Tom asked Sara which card she had liked the best.

Sara said that she liked the one with the rose on the front.

Tom asked if she knew who it was from.

Sara said that she didn't know.

Tom asked if she had liked the one with the heart on the front.

Sara asked how Tom knew that she had got one with a heart on the front.

Tom said surely Sara had recognised his handwriting.

5 *Suggested answer*

A Simon: There's a good job advertised in the paper. I think you should apply.

You: OK, I'll think about it but I'm not sure it's the right job for me.

Simon: Come on, be confident. I really think you ought to send off an application.

You: OK, then, I will.

B You: Simon, they've invited me to go for an interview. I'll phone you afterwards.

Simon: Remember to smile and appear relaxed. What are you going to wear?

You: I'll wear/I'm going to wear my navy suit. I hope they don't notice a button is missing.

Simon: Why don't you sew one on before you go?

C You: Hi, Simon. It went really well. They offered me the job. I'll have to move to Glasgow.

Simon: Good, I'm glad. I'd better tell/warn you that you'll feel lonely/you might feel lonely at first, but don't worry, I'll come and visit you very soon.

Lessons 26–30

VOCABULARY

1   1 beef, pork   2 goulash, spaghetti   3 hamburger, hot dog   4 crab, lobster, mussel, oyster   5 potato, pepper, onion, corn   6 avocado, tomato   7 bake, boil, fry, grill, roast

2   1 thunderstorm   2 island   3 swimming pool   4 afraid   5 beach   6 wreck

3   1 wine   2 steel   3 wool   4 china   5 milk

4   1 T   2 T   3 F   4 T   5 F   6 T   7 T   8 T   9 F   10 T

5   1 plumber – gardener   2 path – sink/drain   3 overflowing – repairing   4 dripping – peeling   5 carpenter – electrician   6 scratched – stained

6   1 d   2 c   3 b   4 e   5 f   6 a

7   1 football, skiing   2 tennis   3 boxing   4 football, boxing, tennis, basketball   5 football, boxing, tennis, basketball

8   1 difficult   2 important/essential   3 impossible   4 dangerous   5 expensive   6 interesting   7 essential/important   8 easy   9 unnecessary

9   2 Going to a party   3 Going for an interview   4 Borrowing something   5 Getting married   6 Using a washing machine   7 Serving a meal   8 Taking pills/medicine.

10   1 had/gave, did   2 give, take   3 done, have   4 have, give   5 take, do   6 give, do   7 do, take   8 do, make   9 make, take   10 give, did

GRAMMAR

1   2 Who is the person you would most like/you most want to meet?

3 What is the event that has most changed your life?

4 What's the book you've enjoyed most?

5 What's the name of the place where you grew up?

6 Tell me about the people/who are the people whose dog you rescued(?).

7 What's the most expensive thing you've ever bought?

2   Cricket is played in India.

The radio was invented by Marconi.

Whisky is made in Scotland.

Diana's funeral was watched by millions of viewers.

Languages have been spoken for thousands of years.

Rockets have been launched from Cape Canaveral since 1961.

Their wedding will be celebrated next year.

3   2 The fruit needs washing.

3 The meat needs chopping into small pieces.

4 The pie needs putting into the oven.

5 Some cutlery needs putting on the table.

6 The wine needs pouring into the glasses.

4   3 Make sure you send out the invitations.

4 Make sure you send all the guests a map.

5 Don't forget to make the bridesmaids' dresses/have the bridesmaids' dresses made.

6 Make sure you have my father's suit cleaned.

7 Don't forget to order a cake.

8 Make sure you book the reception.

5   2 I'm having the bath replaced on Tuesday.

3 I'm having the bedrooms painted on Wednesday.

4 I'm having the door fixed on Thursday.

5 I'm having the light-switch mended on Friday.

6 I'm having the grass cut on Saturday.

7 I'm cleaning the whole house myself on Sunday.

6   2 we do it ourselves.

3 they decorated it themselves.

4 she fixed it herself.

5 he does the washing up himself.

6 you must type it yourself.

7   JANET   2 They don't let me smoke.

3 They make me do exercises.

4 They make me take cold showers.

5 They don't let me stay up late.

KATIE   2 They let me help with the chickens.

3 They let me feed the pigs.

4 They let me milk the cows.

Lessons 31–35

VOCABULARY

1   2 a ticket, a paperback novel and a handkerchief

3 a mirror, a diary and a ballpoint pen

4 an alarm clock, a business card and a credit card

5 an umbrella, some cash and a penknife

6 a notebook, a cheque book and an address book

7 keys, two opera tickets and a mobile phone

2   1 light bulb   2 insect repellent   3 neck pillow   4 door wedge   5 ear plugs   6 torch

3   1 boot – bonnet   2 dashboard – windscreen   3 indicator – speedometer   4 exhaust – fuel gauge   5 bonnet – boot   6 accelerator – clutch   7 steering – spare   8 pedal – jack   9 rear light – gear lever   10 brake – bumper

4   1 c   2 b   3 d   4 a   5 b   6 d

5   1 b   2 a   3 c   4 a   5 b   6 c

6   1 economic   2 environmental   3 encourage   4 appreciate   5 cultural   6 cheap   7 hospitable   8 public   9 sights   10 out of season   11 avoid   12 embarrassed   13 caused   14 experience

7   VERB - NOUN: protect, protection; exhaust, exhaustion; frustrate, frustration; speculate, speculation; survive, survival; arrive, arrival

NOUN - ADJECTIVE: fame, famous; mystery, mysterious; prominence, prominent; indifference, indifferent;

NOUN - NOUN: feminist, feminism; lecturer, lecture; correspondent, correspondence

GRAMMAR

1   2 If the fuel gauge is on the red mark, stop at the next petrol station.

3 If the steering wheel looks well-used, I always check the mileage.

4 Join a motoring organisation in case your car breaks down a long way from home.

5 I switch on the fog lights if it's foggy.

6 A light on the dashboard comes on if you forget to fasten your seat belt.

7 I take my cheque book in case the petrol stations don't take credit cards.

2   2 By following a map you can avoid the motorways.

3 If you use a neck pillow you can easily sleep on trains and planes.

4 By using insect repellent you can protect yourself from mosquito bites.

5 By holding your breath and counting to ten slowly you can stop hiccups.

6 To tell if an egg is hard-boiled, spin it.

3   1 They are due to arrive at half past seven. As soon as they phone me from the station, I will start cooking.

2 It is likely to take us five days to reach the start of the river. After we have set up camp we can explore the jungle.

3 He is expected to pass all his exams. When we hear/have heard his results we will confirm his appointment.

4 When you hear strange noises in the night you must keep absolutely still. Any sound is likely to be made by a bear.

5 After the doctor has examined him she will tell us what is wrong with him. He is likely to stay in hospital for at least a week.

6 Leave the room as soon as she falls asleep. My husband and I are due to return at about midnight.

4   1 If you leave your phone number, I'll ring you when it's ready.

2 If she grows any taller, she'll be the tallest in the family.

3 If I don't pass this time, I'll give up trying.

4 If he continues driving at this speed, he'll beat the record.

5 If you give me the recipe, I'll make it at home.

6 If you embarrass your hosts, they won't invite you again.

7 If the taxi doesn't come soon, I'll miss my flight.

5   2 They had eaten as much as they could, and so they called the waiter to ask for the bill.

3 She told me she had seen a ghost.

4 He fell into the water and drowned.

5 The policeman told her she had been speeding.

6 When we had been travelling for a week, I began to feel sick.

7 He had been waiting since January and it was now October.

8 She didn't stop polishing until she had made everything shiny.

9 The spy was recognised at the airport because he had forgotten to put on his glasses.

10 She decided to join a feminist movement because she felt she had always lived in a male-dominated world.

Lessons 36–40

VOCABULARY

1   **GB**: nursery school, primary school, middle school, comprehensive school, university
    US: kindergarten, elementary school, junior high school, high school, university.

2   1 history   2 physics   3 law   4 geography   5 chemistry
    6 maths   7 medicine   8 economics   9 philosophy
    10 languages

3   1 season   2 mall   3 junk   4 lobby   5 stride   6 routine

4   1 b   2 c   3 b   4 a   5 b   6 a

5   1 couldn't put my finger on it.
    2 'm all fingers and thumbs.
    3 can't make head nor tail of it.
    4 've put my foot in it.
    5 'm all ears.

6   2 unaware – oblivious   3 fellow – guy   4 pay no attention to – ignore   5 cover – lid   6 friend – companion
    7 strange – odd

7   1 jealous   2 asked me out   3 in   4 got   5 broke
    6 get married   7 went out   8 get involved   9 fell out
    10 got engaged

    1-d   2-b   3-g   4-i   5-e   6-j   7-a   8-h   9-c   10-f

8   2 fish   3 stubborn   4 working   5 greedy   6 log   7 thick
    8 horse

GRAMMAR

1   2 If you went to Hong Kong, you would see some spectacular scenery.

3 If you wore flip-flops to school, the headmaster would send you home.

4 If you respected your best friend's feelings, you would not flirt with her boyfriend.

5 If supply exceeded demand, the company would lose a lot of money.

6 If you passed your exams, you would be able to go to university.

7 If I saw a ghost, I would think I was going crazy.

2   *Suggested answers*

1 If I were you, I would have the tiles replaced.

2 I think you should see a doctor.

3 If I were you, I would throw it away.

4 I think you should say yes.

5 If I were you, I would keep the door locked.

3   2 You must have/shouldn't have gone at peak season.

3 You shouldn't have sealed it.

4 You can't have had any trouble at customs.

5 You should have brought your bikini.

6 You must have been cramped.

7 He can't have put in/been wearing his hearing-aid.

8 You shouldn't have been wearing those high heels.

9 They can't have been speaking English.

4   1 They may have gone sightseeing.

2 They might have met in the drugstore.

3 They could have disappeared/gone down a side street.

4 You may have thrown them away.

5 She might have felt threatened.

6 They could have got a taxi.

6   2 If they had used corporal punishment at my school, I would have been beaten quite often.

3 If they had hurried, they wouldn't have missed the train.

4 If you had asked out my best friend, she would have accepted.

5 If you hadn't lost your umbrella, it would have kept us dry.

6 If you had told me, I wouldn't have believed you.

7 If you had asked him for his telephone number, he would have thought that you fancied him.

8 If I had had enough money, I would have bought a silk dress in Hong Kong.

7   2 I would have called for help.

3 I had a dog for company.

4 I will/could lose my job.

5 I hadn't lost my temper.

6 I would be in London my midnight.

7 I get/will get sunburn.

8 I will have some beautiful flowers in spring.

9 I would like to have been/I would have like to be a pilot.

10 I wouldn't have seen a giant tortoise.